COMPRESSED AIR SYSTEMS
A GUIDEBOOK ON ENERGY AND COST SAVINGS

Second Edition

COMPRESSED AIR SYSTEMS

A GUIDEBOOK ON ENERGY AND COST SAVINGS

EDWIN M. TALBOTT

Second Edition

Published by
THE FAIRMONT PRESS, INC.
700 Indian Trail
Lilburn, GA 30247

Library of Congress Cataloging-in-Publication Data

Talbott, E. M.
    Compressed air systems: a guidebook on energy and cost savings /
E. M. Talbott. --2nd ed.
    p. cm.
Includes bibliographical references and index.
    ISBN 0-88173-145-5
    1. Compressed air--Energy conservation. I. Title

TJ985.T35   1992   621.5'1--dc20   92-29641
                                                        CIP

*Compressed Air Systems: A Guidebook On Energy And Cost Savings /
E. M. Talbott, Second Edition.*

Published by The Fairmont Press, Inc.
700 Indian Trail
Lilburn, GA 30247

Printed in the United States of America

10 9 8 7 6 5 4 3 2 1

ISBN 0-88173-145-5   FP

ISBN 0-13-175852-7   PH

While every effort is made to provide dependable information, the publisher, authors, and
editors cannot be held responsible for any errors or omissions.

Distributed by PTR Prentice-Hall, Inc.
A Simon & Schuster Company
Englewood Cliffs, NJ 07632

Prentice-Hall International (UK) Limited, London
Prentice-Hall of Australia Pty. Limited, Sydney
Prentice-Hall Canada Inc., Toronto
Prentice-Hall Hispanoamericana, S.A., Mexico
Prentice-Hall of India Private Limited, New Delhi
Prentice-Hall of Japan, Inc., Tokyo
Simon & Schuster Asia Pte. Ltd., Singapore
Editora Prentice-Hall do Brasil, Ltda., Rio de Janeiro

TABLE OF CONTENTS

PREFACE

This book is an updated re-editing of materials reported in U.S. Government Report DOE/CS/40520-T2 (DE84008802), *Compressed Air Systems,* and references U.S. Government Reports NTIS SAN-1731-T1 and SAN 1731-T2, *Comparative Study of the Energy Characteristics of Powered Hand Tools.*

This previous work by United States Department of Energy (DOE) contractors, most especially the authors of this report, while performing another DOE contract, have led to firsthand, factory-level experience with compressed air systems and the energy factors associated with them. These observations of factory operations, and how compressed air is being used, were made by objective engineers having no economic interest either in the factory operation or in the marketing of compressed air equipment. It was noted that:

1. Compressed air is a popular and highly productive utility that is growing in use, and with good reason.

2. In most factories, there were many opportunities for energy conservation in the generation, distribution, and use of compressed air.

3. In some factories, there was found to be an outrageous waste of energy associated with compressed air systems.

4. Often, compressed air equipment selection, distribution system layout, and final application were made by medium and lower levels of management. Some of these decisions have had major economic consequences without a consideration

of lifetime cost factors and, in some cases, without good engineering practice.

In some factories, it was found that the hissing of air leaks was audible even above the din of a manufacturing operation. In others, pressure drops were found to be on the order of 40 percent of the original compressor discharge pressure. Compressors were being added to systems for the reason that "the pressure is too low at the work stations" without even a cursory analysis of the reason for the low pressure.

The cost of this to the nation in lost energy and to the participating factories in lost dollars is significant. Over the lifetime of the compressed air system, energy is the greatest single cost, in most cases exceeding 30 percent of the total cost involved. If, as in some of the worst cases, 40 to 50 percent of this is being wasted, or, even in some of the more favorable cases where only 10 to 20 percent is being wasted, it is easy to see that the penalty for the improper management of compressed air is a high one indeed. Such is the motivation behind this publication.

Industry sources have estimated that the total connected horsepower of our factory compressed air systems exceeds 17 million. This presents a worthy target for the application of energy conservation technologies. Many energy-conscious engineers exposed to factory environments believe that at least 10 percent and perhaps 20 to 35 percent of this could be saved.

This guidebook seeks to increase the awareness of the compressed air equipment designers, system engineers and air users to the opportunities for conserving energy, by suggesting methodology and offering approaches to energy and cost analysis, it is expected to motivate conservation through better system design and management. It was decided that such a guidebook, based upon investigations sponsored by a United States Government agency dedicated to energy conservation, technical objectivity, and commercial impartiality, will make an authoritative and persuasive statement.

It has been the intention of the authors that the book be factual and understandable to working level engineers and supervisors. It is intended that "the results be seen on the factory floor." Therefore, a style somewhat less formal than that of most technical reports has been adopted. And yet, it is believed that the significant data and methodology have been covered.

Special appreciation must be expressed to the many members of the compressed air industry for their generous assistance with this publication. Companies, industrial associations and individuals have contributed thought, suggestions and constructive criticism to the early drafts of this work in an effort to make this a valuable aid to compressed air users.

1.0 INTRODUCTION

1.1 Purpose

The purpose of the guidebook is to assist industry in the efficient use of compressed air. Compressed air is one of the most important aids to the manufacturing and process industries of any modern nation. It frequently is known as the fourth utility, and it has proved to be an important tool for increasing productivity when applied properly.

The generation of compressed air can be one of the more energy consumptive activities taking place in an industrial operation. It has been estimated that throughout the United States the energy used just to compress air in industry (not including small and portable compressors) exceeds one half quadrillion BTU per year. The use is increasing as our industrial sector grows. Therefore, it is in the national interest that something as important and useful as compressed air be generated and used with efficiency and economy wherever possible. Not only will the result be energy saved, but it also will be dollars saved. It is the intention of this guidebook to outline ways in which compressed air systems can be better designed, laid out, procured, operated and maintained in order to achieve greater energy efficiency and dollar economy.

1.2 Intended Audience

This guidebook is directed to compressed air users, compressed air systems operating and maintenance personnel, compressed air system and equipment engineers, and compressed air equipment manufacturers. The technical level is accessible to most persons in these activities.

1.3 Scope

Specific areas of potential are manufacturing facilities of all sizes and types where compressed air is used on a regular, continuous, or near-continuous basis. Compressor systems in industry and manufacturing tend to operated on high duty cycles, service many different types of tools and processes, and have distribution systems covering large areas. Therefore, they offer opportunities for energy loss and hence for energy conservation. This report is directed primarily to such installations.

Some of the discussion and suggestions will, of course, be applicable to the process industries and to the less energy intensive and lower duty cycle usages such as service stations and construction. In the case of process industries, compressed air more often is treated as an ingredient of the process rather than as a utility. The energy costs incident to compressed air in these industries is a design parameter which tends to be monitored more carefully. Typical construction compressors are portable, operated only a few hundred hours per year and serve local sites. Service stations and other portable compressor applications tend to be small and to operate at low duty cycles. Some of this guidebook, such as the sections on maintenance (4.2.7 and 4.3) will apply to these systems also.

But the prime targets for the energy conservation measures contemplated in this book are the wide variety of manufacturing facilities of many sizes and categories. The compressed air systems and equipment to be discussed will be the types most typically found in such installations. It is not intended to be a complete engineering guide to air compression; the reader is presumed to have a rudimentary knowledge of air compression systems and equipment, the basic units of power and compressed air flow, and normal factory practices.

Within these boundaries then, the book discusses the configurations and ingredients of compressed air systems as they relate to the energy used by the systems and the possible economic impact of the energy requirements. Various approaches are suggested for reducing the energy used without impairing the utility of the systems. A bibliography is included to assist those who wish to study these systems in greater detail.

1.4 Content

The main body of the report begins with a description of typical compressed air systems which are found in manufacturing operations. This includes some diagrams of such systems. Next, Section 3 describes some of the components and subsystems typically found in industry and some of the engineering practices normal to most compressed air installations. This is to establish the fundamental engineering frame-of-reference for the parts of the report which follow. Readers well versed in compressed air systems may wish to bypass or scan Sections 2 and 3.

The main thrust of this report is included in Sections 4 through 6, and the appendices. These describe specific energy conservation practice, recommendations, experience, and economic analyses.

1.5 Other Considerations

Finally, it should be noted that in emphasizing energy conservation this guidebook addresses part of a large subject. The factory manager must, in the final analysis, be responsible for his choices and operations and thus must consider all aspects of compressed air systems such as acoustic effects, safety, and bottom-line economics. In this regard, it would not be inappropriate to question the use of compressed air at all in certain applications. It is not one of the most efficient ways known to transfer and transform power. For example, one government study showed that pneumatic powered hand tools use 5 to 20 times as much power as electric ones to perform the same tasks, if all system factors are considered. There are many installations of compressed air systems that might have been better left to electrical or hydraulic systems. But overall factory economics, including labor costs, must determine this choice and such analyses are beyond the scope of this book.

1.6 How to Use This Guidebook

The sections of this book which may be of greatest interest to a particular user will be determined by the nature of that reader's concerns with compressed air. Some parts of the book provide an overview of typical industrial systems, others will be of greater assistance to system designers, others to maintenance supervisors, and others to operations and energy management personnel.

Appendix B lists the abbreviations used most frequently. Table 1-1, following, lists specific areas of application, with recommended sections. Reference to this table can speed answers to specific problems.

Table 1-1 — GUIDEBOOK APPLICATION

Type of Application	Considerations	Sections of Primary Interest	Sections for Background Information
New System Design	Air demand, pressure, quality; location of work stations; life-time cost; initial cost; growth potential.	3.0, 4.0, 5.0 Appendix A	2.0, Appendix A
System Modification	Same as above plus constraints introduced by existing system.	3.0, 4.0, 5.0 Appendix A	Appendix B
Compressor Replace-ment or Addition	Increased/decreased capacity; growth poten-tial; distribution system capacity; life-time costs; initial costs.	2.0, 3.0, 4.0, 5.1	Appendix B

Table 1-1 (*Continued*)

Type of Application	Considerations	Sections of Primary Interest	Sections for Background Information
Addition to Distribution System	Location of new sections; adequacy of existing headers; pressure drop; installation and ΔP costs.	3.10, 4.1.9, 5.0 Appendix A	3.0, 4.0 Appendix B
System Maintenance	Pressure drop; leakage; corrosion; maintenance costs.	2.0, 4.3, 5.2.3	Appendix B
System Operations	Plant efficiency; air availability; up-time; operating costs; adequate air pressure and quality.	2.0, 3.0, 4.2, 5.2	4.0, 5.0 Appendix B

Table 1-1 (*Concluded*)

Type of Application	Considerations	Sections of Primary Interest	Sections for Background Information
System Cost Analysis	Operations requirements; equipment costs; all other costs; cost of money.	5.2	3.0, 4.0 Appendix B
Pressure Too Low	Operations requirements; system capacity; system condition.	4.1.2, 4.2, 4.3 5.1 Appendix A	3.0, 4.0, Appendix B
Air Too Wet	Operations requirements; system capability; cost of drying.	13.7, 3.8, 3.9, 4.1.2,4.1.7, 5.0	Appendix B

2.0   GENERIC SYSTEM DESCRIPTION

Most compressed air systems consist of the following three major subsystems:

- compressors, with drives controls, inter-cooling, compressor cooling, waste heat recovery, and air inlet filtration

- conditioning equipment, consisting of aftercoolers, receivers, separators, traps, (also frequently called drain traps or drains), filters, and air dryers

- air distribution subsystems, including main trunk lines, drops to specific usage, valving, additional filters and traps (drains), air hoses, possible supplementary air conditioning equipment, connectors, and often pressure regulators and lubricators.

Not all compressed air systems include all of the equipment listed above. Furthermore, there are many ways to compress air, condition it and distribute it to its point of end use. Therefore, there is no standard system which would fulfill every need.

This book is not meant to be a comprehensive analysis of all types of air compression systems that can be designed. Instead, it will concentrate on those most often found in industry and on the energy relationships in those systems.

It will explore four compressor types: single acting recipro-cating, rotary positive displacement, double acting reciprocating, and centrifugal compressors. There are others which are not used, or expected to be used, as frequently. Similarly, the types of air distribution systems and conditioning equipment discussed are those most likely to be encountered, and there will be some variation from factory to factory. More information about air compression systems can be found by reference to the materials in the bibliography included.

With these observations in mind, then, typical systems which will be the subject of this discussion can be found in Figures 2-1 through 2-6. References to these will be made from time to time.

Figure 2-1 illustrates a typical industrial compressed air system. Figure 2-2 shows a representative single acting recipro-cating compressor installation. Figures 2-3 and 2-4 illustrate, respectively, oil cooled and dry rotary positive displacement systems. Figure 2-5 typifies a double acting reciprocating com-pressor and Figure 2-6 shows a similar system incorporating a centrifugal compressor.

Regarding Figures 2-2, and 2-3, the compressors are shown mounted upon the reservoir. This tends to be standard practice only for the smaller systems, i.e., 30 horsepower and under. When the user has control of his reservoir location, he may wish to locate it downstream from separators and dryers, since the latter are more effective and more efficient when flow rates are more uniform.

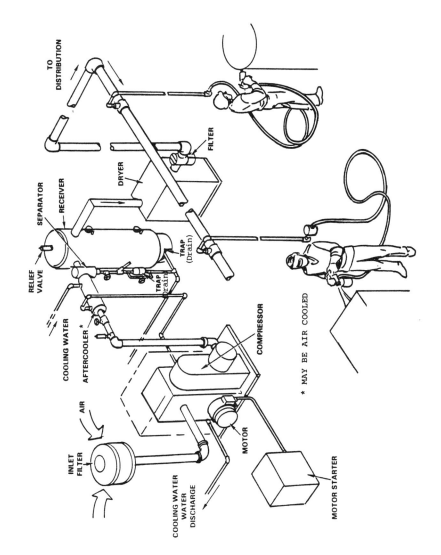

Figure 2-1 An Industrial Compressed Air System

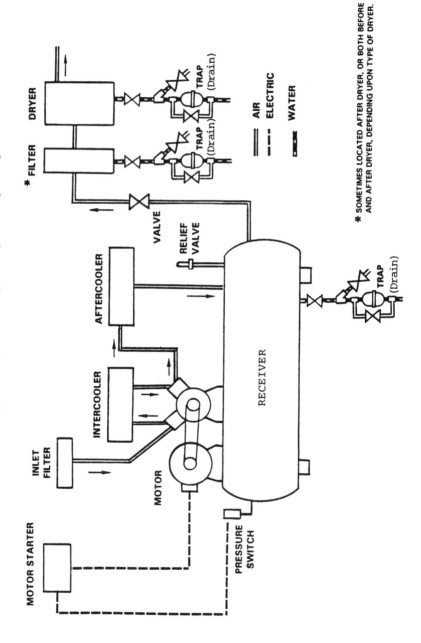

Figure 2-2 Single Acting, Two Stage Reciprocating Compressor

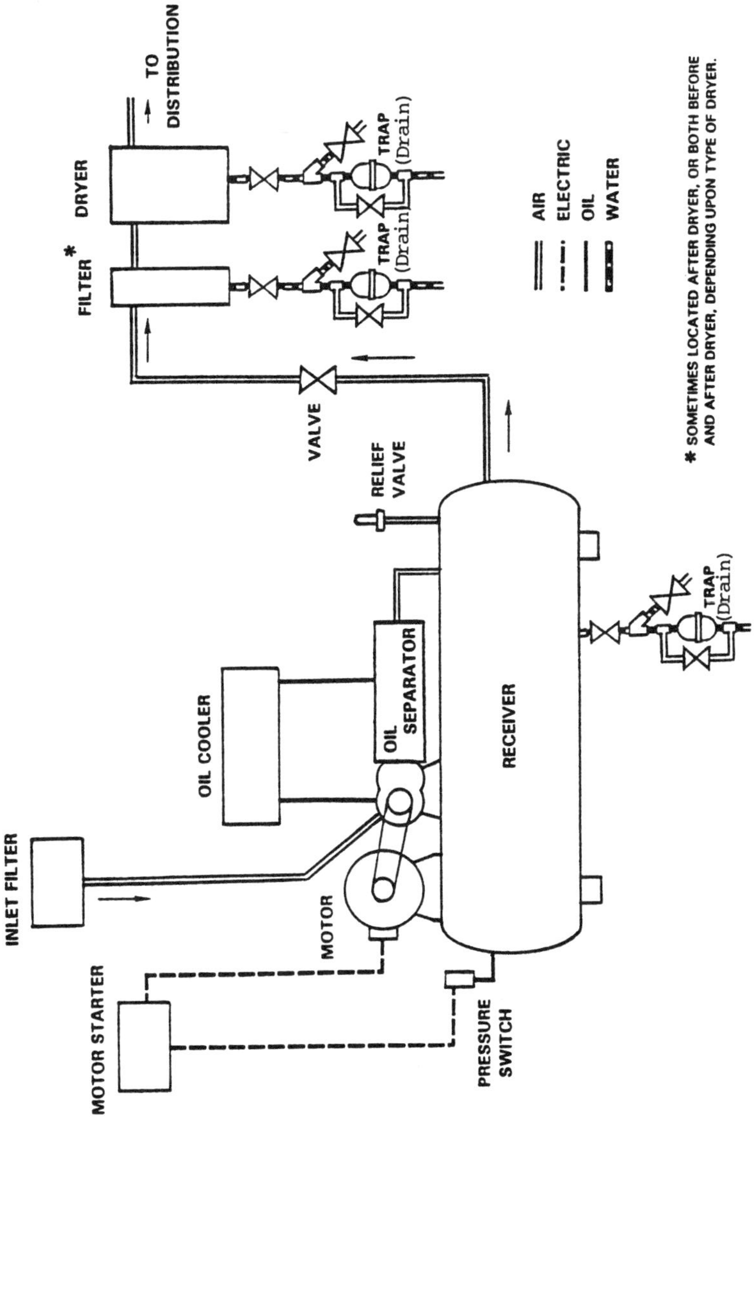

Figure 2-3   Oil Flooded Rotary Positive Displacement Compressor

TO DISTRIBUTION

DRYER

FILTER *

VALVE

RELIEF VALVE

OIL COOLER

OIL SEPARATOR

RECEIVER

INLET FILTER

MOTOR

MOTOR STARTER

PRESSURE SWITCH

TRAP (Drain)

AIR
ELECTRIC
OIL
WATER

* SOMETIMES LOCATED AFTER DRYER, OR BOTH BEFORE AND AFTER DRYER, DEPENDING UPON TYPE OF DRYER.

Figure 2-4 Dry Rotary Positive Displacement Compressor

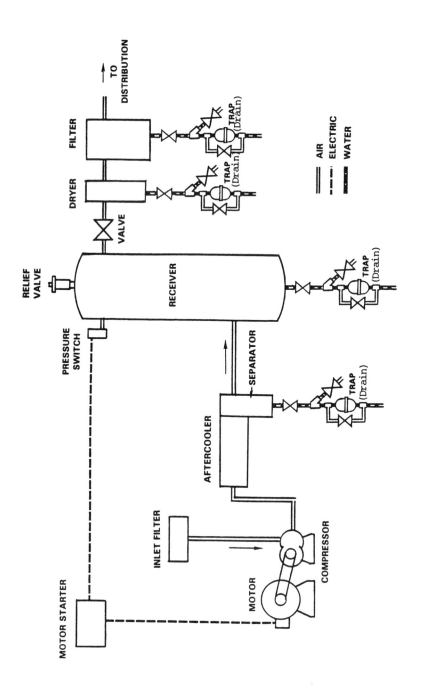

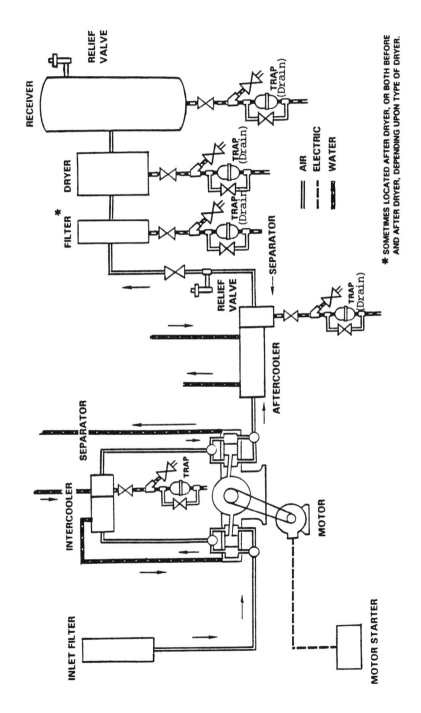

Figure 2-5   Double Acting, Two Stage Reciprocating Compressor

Figure 2-6 Centrifugal Compressor

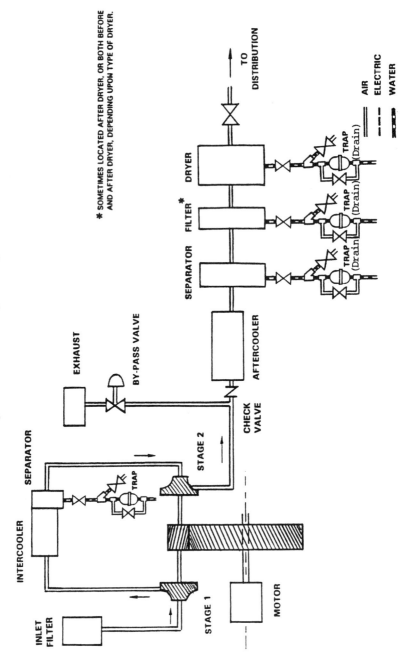

3.0 Components of Compressed Air Systems

3.1 Compressors

While many types of air compressors are manufactured currently, industrial plant air systems are dominated by four types of machinery. These are classified as either positive displacement or dynamic. Positive displacement compressors mechanically displace a fixed volume of air into a reduced volume. Dynamic compressors mechanically impart a velocity to the air through the use of impellers rotating at high speed in an enclosed housing. The air is forced into a progressively reduced volume. The performance characteristics of the two types are markedly different. The positive displacement compressor will deliver a nearly constant volume when operated at a fixed speed while the discharge pressure is determined by the system load conditions. The dynamic compressor volumetric flow will vary inversely with the differential pressure across the compressor.

3.1.1 Types

The various types in each class are:

Positive displacement
Reciprocating: single acting, double acting
Rotary: helical screw, sliding vane, rotary lobe

Dynamic
Centrifugal
Axial (rarely used)

Compressors also are characterized by the quality of the air produced: the compressors may be lubricated or oil free. Each of the positive displacement types is available in an oil free or lubricated design. Centrifugal compressors are inherently oil free, especially when designed to isolate bearing lubrication from the airstream.

Plant air compressors can vary greatly in size from less than 5 to more than 11,000 hp. Here, this broad range has been divided into four categories:

A   5-30 hp
B   30-150 hp
C   150-300 hp
D   300 hp

The combination of four compressor types and four size ranges can be summarized in the matrix shown in Table 3-1. Each of the compressor types is discussed individually in the following sections.

3.1.1.1 Reciprocating Compressors—Single Acting

Reciprocating compressors have been the most widely used for industrial plant air systems. The two major types are single acting and double acting, both of which are available as one or two stage compressors.

The single acting cylinder performs compression on one side of the piston during one direction of the power stroke. Two stage compressors reach the final output pressure in two separate compression cycles, or stages, in series.

3.1.1.2 Reciprocating Compressors—Double Acting

The double acting compressor is configured to provide a compression stroke as the piston moves in either direction. This is accomplished by mounting a crosshead on the crank arm which is then connected to a double acting piston by a piston rod. Distance pieces connect the cylinders to the crank case. They are

Table 3-1—Matrix of Available Plant Air Compressor Types Versus Size—100 psig Operation

Size,hp (cfm)	Reciprocating		Rotary Positive Displacement	Dynamic
	Single Acting	Double Acting		
5-30 (20-120)	One & Two Stage	Normally One Stage	Usually Rotary Screw	Not normally manufactured
30-150 (100-750)	One & Two Stage	Normally Two Stage	Usually Rotary Screw or Lobe Type	Not normally manufactured
150-300 (600-1600)	One & Two Stage	Normally Two Stage	Usually Rotary Screw	Usually Centrifugal
>300 (>1300)	Not normally manufactured	Normally Two Stage	Usually Rotary Screw	Usually Centrifugal

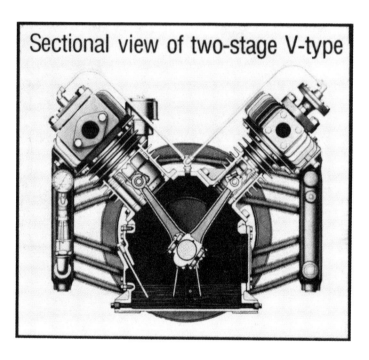

Figure 3-1 Single Acting Reciprocating Compressor
Courtesy: Gardner-Denver Company

sealed to prevent mixing of crank shaft lubricant with the air, but vented so as to prevent pressure buildup.

3.1.1.3 Rotary Positive Displacement

The double helical screw compressor was introduced commercially in the 1960's and has become increasingly popular for industrial plant air applications. Some of the advantages of this

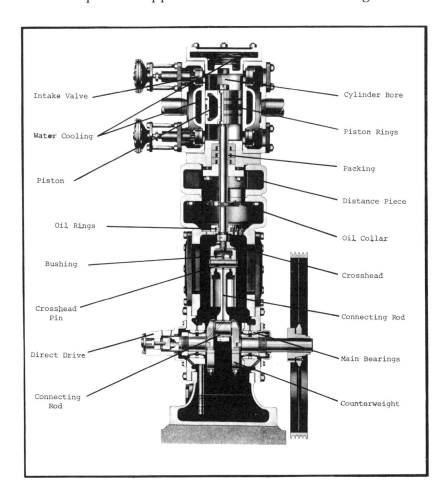

Figure 3-2   Double Acting Reciprocating Compressor, Single Stage
Courtesy: Worthington Compressors, Inc.

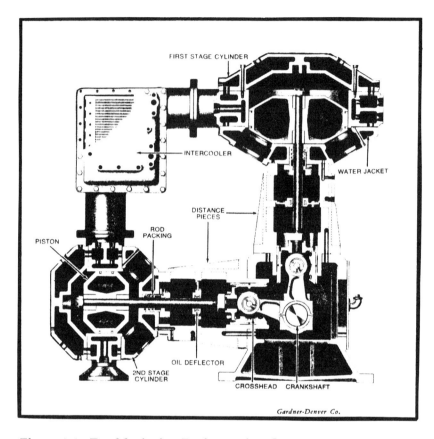

Figure 3-3   Double Acting Reciprocating Compressors, Two Stage
Courtesy Gardner-Denver Co.

type are lower initial cost, lower installation and maintenance costs, reduced vibration, smaller size, and delivery of uniform, minimally pulsing air flow. The double helical screw units are available in oil flooded and nonlubricated designs, the lubricated ones being more efficient and less costly to purchase and operate.

The oil flooded design is illustrated in Figure 3-4A. Air is compressed by the two rotating, intermeshing rotors. The action of the rotary screw can be compared to a reciprocating compressor. Each of the slots in the female rotor acts as compressor cylinder while the point of contact between the two rotors acts as a piston. The air is forced along the helical volume around the female rotor toward the discharge end plate. The volume behind the point of contact is drawing air in through the intake port. The cylinder is sprayed constantly with oil through orifices in the cylinder wall. The oil provides a seal between the rotors and cylinder, lubricates the rotor and absorbs most of the heat of compression. Compression ratios of 7 to 8:1 are typical with compression to 100 psi performed in a single stage. Newer designs of rotary screw compressors have modified the cross section of the rotors to achieve more effective sealing characteristics. This has resulted in higher compression efficiencies, so that the rotary screw has become nearly as efficient as a single-stage, double acting reciprocating compressor.

The discharge air temperature typically is 90-100°F above ambient (if air cooled), since the lubricating oil acts as a coolant. An oil to air or oil to water heat exchanger is required to remove this heat from the lubricant. Some portion of the lubricating oil inevitably is carried over in the air. A conventional oil removal system can reduce the oil content to 3 to 15 parts per million.

The dry (nonlubricated) screw compressor is similar in design, except that external rotor positioning or timing gears are used to insure that the unlubricated rotors do not contact each other. The rotors are machined to closer tolerances than the oil flooded units and seals must be so designed that lubricants can be isolated from the air stream. Also, since there is no oil seal between the rotors and casing, the operating speed must be high (up to 20,000 rpm) in order to minimize clearance space leakage and hence to achieve reasonable efficiencies. The compression ratio of each stage normally is limited to 3.0 to 3,5, so that two stages are used for 100 to 125 psi plant air systems. This type of compressor delivers nearly oil-free air at high temperature, necessitating after-cooling and, in larger sizes, liquid cooling of the compressor. The noise level is higher than that of the lubricated version, because of the higher speeds, but special noise attenuating enclosures normally are supplied. See Figure 3-4B.

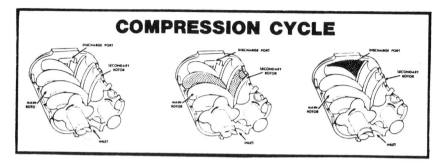

A - Oil Flooded Compression Cycle
Courtesy Gardner-Denver Co.

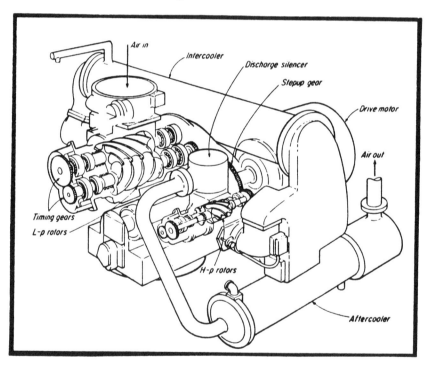

B - Nonlubricated Two Stage Compressor Design
Courtesy Power Magazine *April, 1972*
© 1972 *McGraw-Hill, Inc. All rights reserved*

Figure 3-4   Rotary Screw Positive Displacement Compressors

Rotating lobe compressor designs have undergone recent modifications to improve efficiency and reliability for discharge pressures of 100 to 125 psi. The newer rotating lobe designs have several inherent advantages over the rotary screw. Thrust loading on the main bearings is eliminated and reliability can be increased by using heat shrunk rotors and gears and water cooling. The compressors normally are designed to be nonlubricated. Otherwise, except for a somewhat less smooth outflow, their characteristics are similar to those of rotary screw compressors.

Other types of rotary positive displacement compressors are available, but they are used infrequently in industrial plant air systems. Examples are rotating vane and single screw compressors.

3.1.1.4 Centrifugal Compressors

The centrifugal compressor has seen increasing usage for high volume plant air applications (from 700 to 30,000 cfm) as technological improvements have gradually increased efficiencies. These compressors operate at high speeds, resulting in smaller, smoother, more compact equipment. Plant air systems operating at 100 psi normally require two or more stages of centrifugal compression, all stages normally driven by one large bull gear. Figure 3-5 illustrates a typical three-stage centrifugal compressor.

The basic element of the centrifugal compressor is an impeller, which is mounted on a shaft and positioned within a housing consisting of an inlet, a volute, and a diffuser. The impeller spins at high speed and imparts a velocity to the air. The diffuser surrounds the impeller and acts to convert the kinetic energy of the air into potential energy at a higher pressure level. The volute further reduces the velocity and converts more kinetic energy to potential (pressure) energy.

An important and potentially troublesome characteristic of centrifugal air compressors is that of surge. As the compressed air demand is reduced, the system pressure rises until a critical point

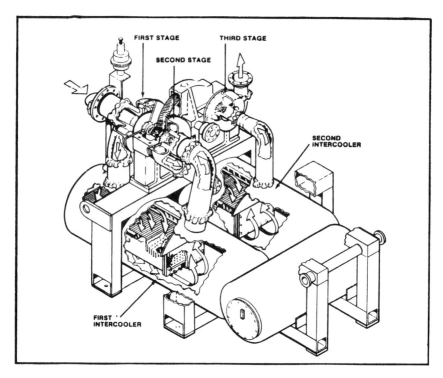

Figure 3-5   Three Stage Centrifugal Compressor
Courtesy Elliott Company, Jeannette, PA

is reached. At this critical pressure, the discharge air flows back into the compressor, immediately reducing the discharge pressure. This causes a surge in flow, again raising the discharge pressure, and inducing a repetition of the flow-back. If there is no interruption, a harmonic continuation of this cycle is established and it can result in destruction of the compressor. It is vital that the compressor be protected from the effects of surge.

3.1.2 Multiple Staging

Generally, all compressor types discussed here will operate more efficiently if designed with multiple stages. In multistage compressors, the final discharge pressure is developed in several

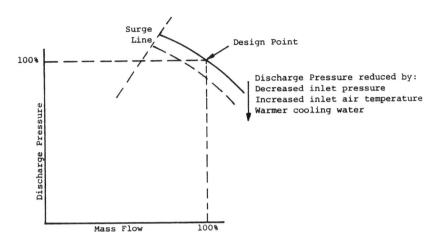

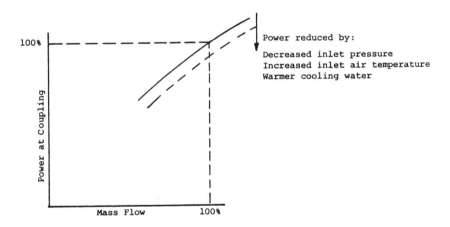

Figure 3-6   Some Characteristics of Centrifugal Compressors
Varigas Research, Inc.

sequential steps, using different compression hardware to connect the output of the first stage to the inlet of the second. The
energy saving is achieved by cooling the air between stages, thus
reducing the air volume and hence the work required to complete
compression to the desired working pressure in the second stage
of compression as shown in the indicator chart for a reciprocating

compressor in Figure 3-7. The cycle CBF'G represents single-stage compression while cycles CAFG and C'A'F'G' represent the combined two-stage compression cycle. Intercooling reduces the temperature and volume of the air between stages, resulting in the work saving represented by the shaded area. This relationship applies to all compressor types.

Another purpose of multistaging is to attain higher pressure levels at air temperatures lower than the safe limit for a single-stage compressor, typically 100 to 125 psi. The cooler compression chambers not only reduce fire hazard, but also decrease thermal stresses on the metal lubricant and decrease the loads and mechanical stresses imposed on compressor components, particularly valves, rings, and bearings.

3.1.3 Compressor Cooling

The process of compression generates much heat. Compressors normally are cooled by either water or air, except that oil flooded rotary positive displacement types use the lubricating oil as a coolant. Normally, air cooling is limited to smaller sizes (less than 150 hp) of reciprocating compressors, or compressors that are designed for intermittent service. Air cooling is accomplished by providing a series of fins along the outer wall of the compressor to increase the surface area. Frequently, the compressor fly wheel is designed to serve as a fan to force additional air across the cylinders.

Water cooling is more effective because of ease of control and greater heat absorption capability per unit volume. Water cooling normally is required for continuous duty compressors. Reciprocating compressor cylinders and cylinder heads include a series of water-filled cavities and passages. The cooling water in non-recirculated systems typically is cooler than the ambient air, so the compressor cooling water normally is routed through the aftercooler or intercooler first so as to reduce thermal stresses and the chance of overcooling the inside surfaces, thus creating condensate inside the compressors.

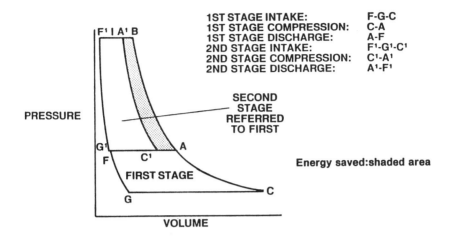

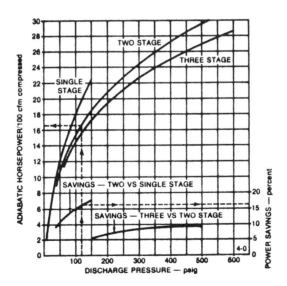

Figure 3-7   Theoretical Energy Savings Possible Through
Multi-Staging Reciprocating Air Compressors
Courtesy Ingersoll-Rand Co.

Two problems arise in water-cooled compressors: supply and disposal of water at reasonable cost. The quantity of water can be reduced if a control system is used to vary the flow in response to the cooling demands of the system. The controller could vary the flow as a function of the water discharge temperature. The control valve always should be on the outlet to maintain pressure in the system so as to prevent pressure reduction induced aeration and resulting air bubbles in coolant passages. Precautions should be taken to deal with solid material that may settle out as a result of water flow throttling.

3.1.4 Compressor Sizing

Two conflicting factors influence the determination of the total compressor capacity needed to supply a system. All constant speed air compressor types are most efficient when operated at full load, i.e., maximum capacity. Different designs have widely variable part load efficiencies, but all efficiencies are maximized at full load. Thus, the most efficient compressor would be sized to handle the average load and would operate normally at full load. On the other hand, the inability to meet peak demands could result in decreased production and hence a much greater total cost. Undersized compressor capability results in reduced system operating pressures, leading to serious reduction in output and efficiency in the operations relying on pneumatics. Figure 3-8 shows the relative effectiveness of an air powered tool as air pressure is reduced from the design value of 90 psi.

The use of multiple compressors with sequential control circuitry is a potential solution to the dilemma by providing a better match of load and operating compressor capacity. Multiple compressors also permit compressor backup for maintenance and repairs. For example, three compressors, each of which is sized for 50 percent of the normal peak load, is one configuration which will offer these advantages. The disadvantages of multiple compressors are that the full load efficiency of smaller compressors generally is less than that of larger ones and that multiple

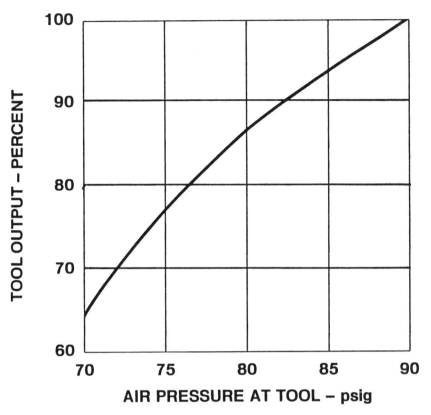

**Figure 3-8 Relative Air Tool Effectiveness as a Function of
Reduced Air Pressure. Tool Designed for 90 psig.**
Varigas Research, Inc.

units are more costly per unit of capacity, to purchase and install.
   If a new compressed air system is being designed, the system
capacity is determined by an analysis such as that tabulated in
Table 3-2, where all known steady air users are identified and
their expected consumption calculated. Air consumption rates of
various tools are available from manufacturers and the Compressed Air and Gas Institute (CAGI) handbook. (See Appendix
A). The load factor represents the percentage of time that a

Table 3-2—Estimate of Air Required for the Steady Load Portion of a New Compressed Air System.

Type of Tool	Location	Number	Load Factor	Air Requirement (scfm) @ 90 psig		
				Per Tool	Total (all on)	Expected Load
Drills	Assembly	10	15	35	350	53
Grinder	Finishing	15	20	50	750	150
Sm. Screw-driver	Assembly	20	10	12	240	24
Lg. Screw-driver	Assembly	10	14	30	300	42
Riveters	Assembly	5	10	35	175	18
Hoist	Cleaning	1	5	35	35	2
Chippers	Cleaning	6	20	30	180	36
						325

Example:

1. Total scfm, all on = number x scfm per tool = 10 x 35 = 350
2. Expected load = load factor x total, all on, scfm = 0.15 x 350 = 53

pneumatic load is operating and the percentage of maximum air consumption under operating conditions. Additional allowances must be made for leakage, typically and hopefully no more than 10 percent, and for any expected future plant growth. This determines the total system requirement. It is normal practice that water-cooled compressors are sized to supply their total system requirements plus 30 percent, and air cooled compressors are oversized by 40 percent. These margins can be cut back if the load estimates are based on specific plant experience rather than estimates.

If an existing system is being enlarged, the existing load factors and the required additional capacity are more easily and accurately measured and determined from operating experience. The proportion of the load handled adequately by the existing compressor system to that of the total of the enlarged system can provide guidance for estimating the additional capacity required. This can be done by monitoring pressure at various locations throughout the plant during peak operating times. Consumption of low usage devices such as air cylinders or feed devices can be estimated by computing their piston displacement and operating frequency. Heavy short run users require special consideration and might benefit from local air storage to maintain pressure during the operating cycle.

After the required system capacity is determined, an initial estimate of compressor capacity can be made. A more thorough procedure for designing a complete system from point of use back to the compressor is presented in Section 4.2.

3.2 Compressor Drivers

Alternating current (ac) electric motors are by far the most common drivers used for industrial plant air compression. It is only in special applications that other drivers are feasible, such as installations that have low cost or surplus steam or natural gas.

3.2.1 Alternating Current Electric Motors

Large electric motors are available in two basic types: induction and synchronous. Nearly all U.S. industrial motors are three phase ac powered. Both the induction and synchronous motors rely upon the production of a revolving magnetic field (rmf) in the field winding as shown in Figure 3-9.

This magnetic field rotates at the line frequency of the power supply in the two pole motor design illustrated (3600 rpm). The rotor normally rotates at a slightly slower speed and the angular amount they are out of phase is called "slip." Four and six pole motor designs also are used, and these revolve at lower speeds (1800 and 1200 rpm).

3.2.1.1 Induction Motors

Induction motors are used in 90 percent of industrial applications, and are designed in two styles: squirrel cage and wire wound rotor. The primary operating differences are the starting torque, current and amount of slip. Only the squirrel cage motor will be discussed, since almost all induction motors are of this type.

The rotor is composed of metal bars in which a current is induced by the cutting of the lines of magnetic force produced by the rmf when there is a non-zero slip angle. This current induced in the rotor sets up a secondary magnetic field around the rotor which interacts with the rmf of the stator to generate a torque which is proportional to rotor current and hence to slip angle. The rotor design can be tailored to meet specific torque and speed requirements. Three standard NEMA design squirrel cage motor characteristics are shown in Figure 3-10. Design B is most frequently used.

An important consideration of ac induction motors is their power factor (pf). When an electrical load is purely resistive, the current through the load is proportional to the voltage, and the current and voltage are in phase as shown in A of Figure 3-11. Power is measured in watts or kilowatts (kW), and is the instan-

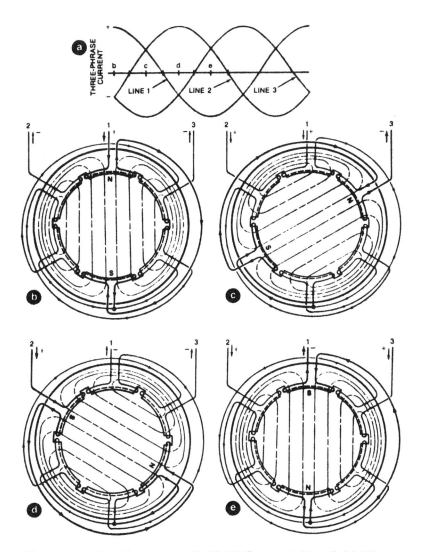

**Figure 3-9   Rotating Magnetic Field Generated In Field, Three
Phase, Two Pole Motor**
Fairbanks, Morse & Co.

Figure 3-10 Typical Speed-Torque Characteristics of NEMA Design, B, C, and D Polyphase Squirrel Cage Induction Motors at Full Line Voltage.

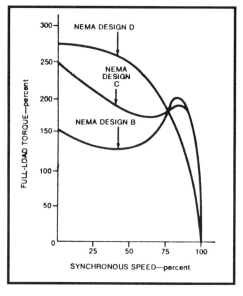

taneous product of current and voltage. Since the ac motor presents a load that is part inductance and part resistance, the current will lag the voltage by a phase angle whose cosine is proportional to the ratio of the active current to the apparent current. Low values of the pf indicate more inductance (vs resistance) in the load and a great phase angle. The practical effect of this is that measured power in kW will be less than the simple product of current and voltage (Volt Amperes or kiloVolt-Amperes, i.e., kVA). These relationships are shown diagrammatically in Figure 3-11, B and C.

The actual line current to the motor, as measured by an ammeter, is the vector sum of the active and reactive currents. The electric utility must deliver this current even though only part of it registers on a Watt-meter. Utilities, therefore, make additional charges based on low inductive power factors. So normally there will be economic benefits to maintaining a low pf. This varies among utility companies.

3.2.1.2 Synchronous Motors

Synchronous motors run at fixed speeds, determined only by the supply line frequency and the number of motor poles. After synchronous speed is reached, motor speed is independent of load, up to the pull-out point, as shown in Figure 3-12. The

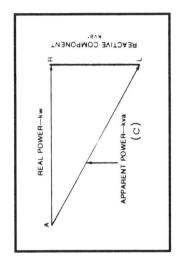

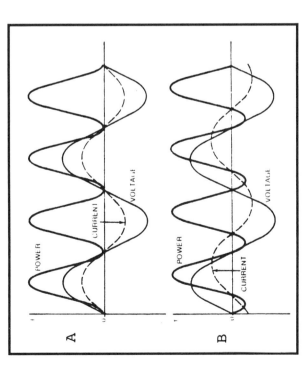

Figure 3-8 Relationship of Current, Voltage, and Power in Resistive and Inductive Loads
Fairbanks, Morse & Co.

primary difference between induction and synchronous motors is that the rotor field coils of the synchronous motor are supplied with a separate dc excitation current rather than relying on slip-generated current.

The synchronous motor normally has a small inductive component on the rotor to enable motor starting since the pure synchronous motor has zero torque at rest. When first started, the inductive winding brings the rotor to 92-95 percent of synchronous speed, whereupon the dc excitation current is supplied to the rotor and the rotor speed increases and locks onto the rmf synchronous speed. The motor maintains this constant speed as long as the pull-out torque is not exceeded. The starting process is somewhat complicated by the need to switch from inductive to dc excited rotor configurations. Any overload controls must also remove the dc excitations.

The motor can be designed with a unity or leading (capacitive) pf. The leading pf can be achieved by over exciting the rotor field at the expense of slightly reduced operating efficiency.

Synchronous motors have several advantages over induction motors including:

• Synchronous motors have a unity or leading pf and can improve industrial plant average pf. Typical motors are 1.0 to 0.9 leading pf with some available to 0.2 pf.

• Synchronous motors have higher electrical efficiencies due to dc excitation of rotors (1-3% higher than same size and speed inductive or dc motor). See Figure 3-13.

• In large sizes (hp greater than rotor speed in rpm), the synchronous motor has lower combined initial and operating cost.

• Constant speed is maintained regardless of load or line voltage.

• The synchronous motor can be designed for low speeds (multi-pole) and direct drive applications.

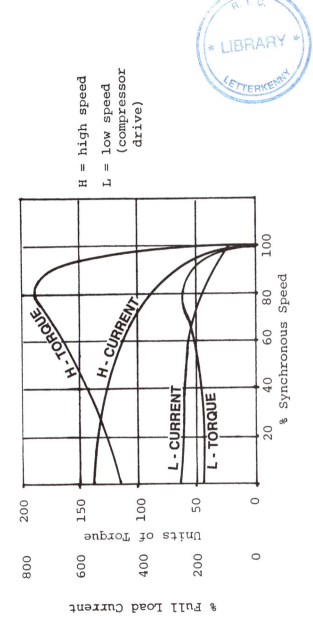

Figure 3-12   Typical Speed/Torque Characteristics of Polyphase Synchronous Motor

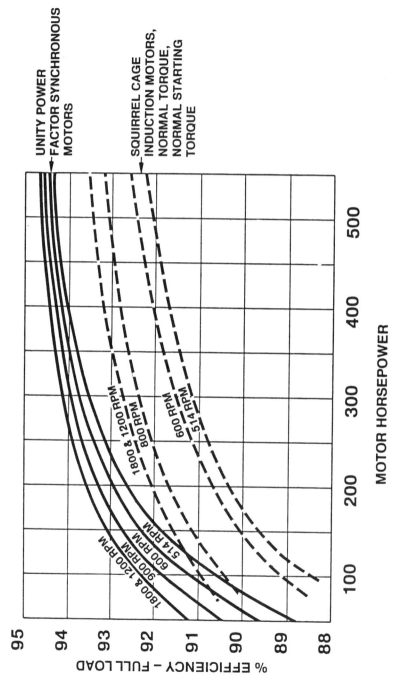

Figure   3-13 Comparison of Electrical Electrical Efficiencies of Synchronous and Squirrel Cage Induction Motors at Full Load as a Function of hp and Design Speed, 60 Hz.

3.2.2 Other Driver Types

Factories which are generating high pressure steam for other reasons may find it economic to operate steam turbine driven compressors, especially if the compressors are dynamic types. Almost always this will involve sizes exceeding 250 hp.

Manufacturers who have use for a large amount of low-grade steam (5-10 psi) on a continuous basis may find it economical to generate the steam at higher pressures and use it for compression and/or other prime mover purposes prior to distribution of the steam for low-pressure application. This is known as co-generation and a great many studies have been made of this subject.

Sometimes the availability of natural gas makes gas-powered compressors economic. Some reciprocating compressors are manufactured for this purpose, as integral engine/compressor machines.

Normally, gas turbine, diesel, or Otto cycle engine power is not economic for stationary, continuous service compressor installations, except in special circumstances.

3.2.3 Driver Connection to Compressors

Small reciprocating compressors most often are belt driven. Larger compressors, reciprocating and rotary positive displacement units normally are flange connected. Some designs of compressors have a shaft directly coupled with the driver. This is typical of large turbine driven dynamic compressors. Speed changing gears in the driving train are seen less frequently, most often for powering centrifugal compressors with electric motors.

3.3 Controls

There is a growing variety of control systems available for compressed air installations. These most often concern electric driver controls and compressor controls. They may be best categorized as:

Electric Driver Controls	Compressor Controls	
Starting	Start/stop	Modified throttling
Variable voltage	Load/unload	Variable speed
Variable speed	Multistep partial load	Multiple compressor
	Throttling	Surge control

3.3.1 Electric Driver Controls

3.3.1.1 Electric Motor Starters

Four types of mechanisms for starting various sizes of three phase induction motors are common and commercially available. The controls for synchronous motors, which are used less frequently, are more complex and should be specifically tailored to each installation by the electrical equipment supplier. They will not be discussed here. The most frequentlyused induction motor starting procedures are:

- Full voltage starting
- Reduced voltage starting
    Star/delta connection
    Resistance
    Auto-transformer
    Solid state voltage reduction

- Variable voltage for starting and reduced load
- Variable speed controllers

Direct full voltage starting is generally limited to small compressors (less than 250 hp). Most often this is accomplished by a conventional motor starter with built in overload protection. Alternate automatic shutdown can be added by connecting the starter relay in series with external switches such as overpressure, coolant temperature, low lubricant pressure, or others. Small units also may use simple, fused three phase switches. Full voltage starting usually has the lowest initial cost and is the preferred method where possible, but it does present some disadvantages:

High inrush current—Starting currents are 5 to 8 times full load current resulting in electrical stresses on the motor and electrical supply system with potentially increased operating costs.

High start-up torque—Starting torques of 1.5 to 4 times full load torque produce stress on motors, shafts, and compressors. During the acceleration cycle, output torque can peak at 3 to 5 times full load torque.

Utilities may not permit full voltage starting of motors due to the high kVA inrush. Some limit starting current to a fixed percent of full load. Others have a demand charge in their rates and this is determined by peak power requirements.

3.3.1.2 Reduced Voltage Starting

Reduced voltage starting can be implemented by any one of several methods, including star/delta connection of three-phase motors, primary resistive or reactive circuits, auto transformers and solid state rms voltage reduction. Reducing the starting voltage directly reduces both the starting current and torque and the problems that can be caused by elevated values of these parameters during start-up. Reduced voltage motor control and protection functions are implemented in a manner similar to that of full voltage starters.

The three-phase star/delta circuit approach initially connects the motor windings to the line in a star or Y configuration which results in a voltage of 0.577 times line voltage across the windings as shown in Figure 3-14. After reaching near synchronous speed, the starting relays are switched to connect the windings across the lines (delta) with full line voltage on each winding. Some regard the star/delta system as not so much a reduced voltage method as a temporary, alternative motor wiring system. It is the most popular method outside the U.S. and has gained widespread acceptance here also.

The series resistance primary starter system is another common and inexpensive approach used for smaller motors. During the starting cycle, resistors are placed in series with the motor, thereby reducing both voltage and current and hence the start-up torque. Multiple stages of relays and resistors can be used for even smoother starts. The series reactor primary starter employs the

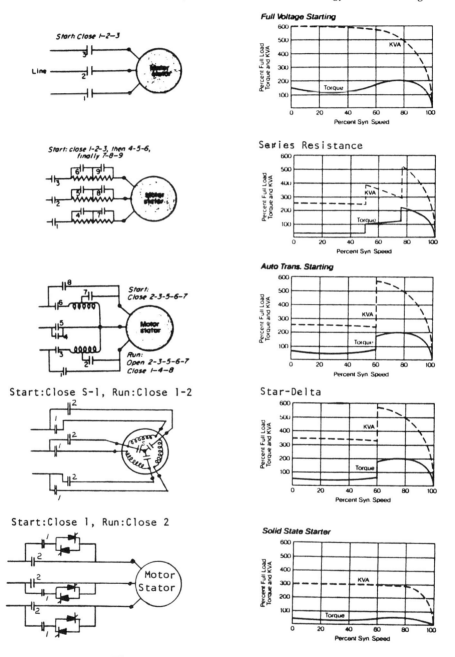

Figure 3-14   Starting Procedures for
Three-Phase Squirrel Cage Induction Motors
Courtesy Vectrol Division, Westinghouse Electric Company

same technique but uses reactors instead of the resistors as the voltage reducers.

Auto transformer starters use Y connected transformers to reduce the starting voltage during start-up. They may be field adjusted to 80, 65, or 50 percent of line voltage and produce starting currents and torques usually at 66, 44 and 26 percent, respectively, of those associated with full voltage starting.

Solid state reduced voltage starters typically use silicon controlled rectifier (scr) devices to reduce the rms voltage to the motor. The voltage delivered to the motor is a chopped ac waveform. These start-up systems typically allow field adjustment of both the starting torque and the time required to ramp up from zero to full voltage. This permits reduction and control of inrush current and mechanical shock to the motor and load. In the simplest form, the scr's are bypassed after the motor reaches running speed. Normally, all standard control features are implemented as part of the package.

3.3.1.3 Variable Voltage for Starting and Reduced Load

Some more recent designs of solid state starters use the same scr's to control the voltage to the motor while operating at full speed under reduced loads. These can reduce motor losses significantly when operating 5 to 50 hp motors at reduced loads. This will be discussed further in Section 4.

3.3.1.4 Variable Speed Controllers

Variable speed controllers are available for ac synchronous and induction motors. These enable continuous speed variations from near 0 to 100 percent, matching compressor output to demand. The controller operates by converting the ac line voltage to dc then generating a variable frequency step function output. Typical inverters generate a six step wave form which approximates an ac sinusoidal wave form. Because of the initial cost and loss of efficiency, these controllers normally are practical only for small reciprocating compressors (less than 25 hp) that frequently operate at reduced loads.

Some variable speed controllers now are being applied to centrifugal compressors, both as the sole control method and in combination with other control approaches.

3.3.2 Compressor Controls

Plant air compressor systems normally are designed to operate at a fixed pressure and to deliver a variable volume. The compressor is sized to deliver the maximum capacity and a control system is employed to reduce the compressor output to match the system demand.

Each compressor type described in Section 3.2 may incorporate any of several different control systems to match the compressor volume and pressure to the demand, and the most popular are described below. All of these controls monitor the system pressure as an instantaneous indicator of the status of the match between compressor output and system demand. Usually the control systems will recognize and be designed to deliver air pressures between a design minimum and a design maximum damped, system pressure. The damping is required to eliminate the effects of pressure pulses produced by most compressors. Pressure differentials of 3 to 25 psi between the minimum and maximum are specified in practice, the actual differential being a function of user requirements. This differential is known as the "control range." The controls described below are summarized in Table 3-3.

3.3.2.1 Type A. Start/Stop and Load/Unload Controls

The simplest control mechanism turns the compressor on or off in response to the system pressure. When the preset system high pressure is reached, the compressor is turned off. When the system pressure falls to the preset minimum, the compressor is turned on.

Delivery capacity is limited to two values, 0 or 100 percent capacity in diagram A of Figure 3-15. The electric motors are

Table 3-3 Commonly Available Compressor Control Techniques For Various Compressor Types and Sizes

Available Compressor Control Techniques by Driver Size

Control Type:	Type A - Two Step		Type B - Multi-Step		Type C - Throttling		Type D - Modified Throttling	
	Start/Stop	Load/Unloaded	Double Acting Unload	Unloaded/Clearance Pockets	Discharge Bypass	Variable Speed	Inlet Valving (Same x/Un-loading)	Modified Throttling Inlet Guide Vane Adjustment
Modulating Range (%FL Capacity)	0,100	0,100	0,50,100	0, 25, 50, 75,100	0-100	20-100	0-100	0-100
Reciprocating Compressor								
Single Acting	W	W,X	X, Y, Z					
Double Acting	W	W,X		Y, Z				
Rotary Compressor								
Helical Screw	W	W, X, Y					W, X, Y, Z	
Lobe	W.	W, X					W,, X, Y, Z	
Vane	W	W, X					W, X, Y, Z	
Centrifugal Compressor					Y, Z		Y, Z	Y, Z

simply started and stopped. This control is normally used only for low duty cycles, low power installations, or applications in which air usage occurs in infrequent cycles. Otherwise motor overheating might occur as a result of frequent cycling.

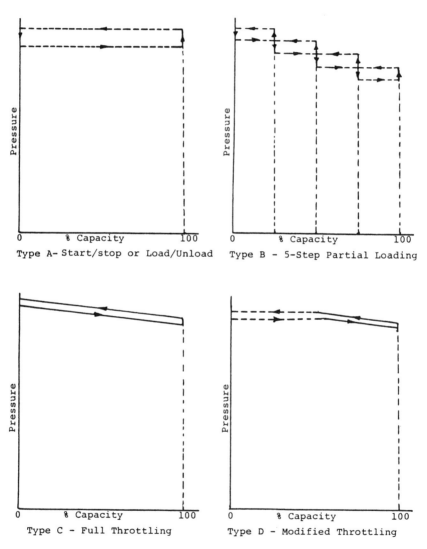

Figure 3-15 Pressure vs Capacity of Compressor Control Procedures

For those applications where stop/start operation is not advisable, the compressor is alternately loaded or unloaded. This is analogous to start/stop control, but the compressor, rather than the driver, is controlled. When the compressor is unloaded, the driver continues to operate, but at a greatly reduced electric demand, and no compressed air is delivered to the system. The cycling can be of short duration without either mechanical or electrical stress.

Each compressor type performs this unloading differently. Reciprocating compressors are unloaded by holding the inlet valves open, preventing air induction and compression. Rotary compressors are partially unloaded by controlling the inlet air and bleeding the discharge pressure to atmosphere so the compressor will not be operating into a high pressure.

3.3.2.2 Type B. Multistep Part Load Control

The output of compressors that can operate in two or more partially loaded conditions can be matched more closely to system demand. Typically, the control system employs multiple pressure switches, operating at (usually) evenly spaced increments within the pressure control range.

Some analog and digital electronic control systems operate within a much smaller pressure control range by resetting the pressure sensor and programming the system to increase or decrease the capacity, one increment at a time. The same can be accomplished using relays, but the electronic systems are state-of-the-art, more reliable, and more trouble free.

An example of multiple step partial load control is illustrated in Figure 3-16. This double acting reciprocating compressor has two control mechanisms on each cylinder. The first, inlet valve depressors, enables either or both of the cylinders to be unloaded by holding the inlet valves open during the compression cycle. No compressed air is delivered and very little energy is used by the defeated cylinder. Using this control independently on both cylinders provides delivery capacities of 100, 50, and 0% of maximum. The second type of control, the use of clearance pockets, permits selectively increasing the effective volumes of the cylinder heads and hence reduces the compression ratio. The clearance pockets normally are sized to reduce the volume of air delivered by 50 percent of the maximum for each cylinder. This control provides the ability to select compressor delivery from each cylinder of either 100 or 50 percent of maximum capacity. Some compressors use two independently controlled clearance pockets for each cylinder, permitting the selection of outputs of 100, 50, and 0 percent of maximum. This offers the same added flexibility as can be achieved by combining a single set of pockets with full unloading (via valve lift) capability.

These types of controls frequently are applied to double acting compressors, resulting in a five step control capability, with delivery capacities of 0, 25, 50, 75, and 100 percent of capacity as shown in Figure 3-16. The card diagrams shown indicate the operating cycles of each cylinder for each of the five steps. See Figure 3-15, Diagram B for the pressure/output relationship.

3.3.2.3 Type C—Throttling

A few compressor types achieve uniform, continuous variation in output by using throttling controls over the complete range from 0 to 100 percent of capacity. Such controls permit close matching of compressor output to system demand.

There are several practical problems in implementing this type of control. Further, the reduced capacity is not normally reflected in a comparable reduction in power required to drive the compressor. As a result, part load operation is less efficient than full load performance. In practice, modified throttling controls, discussed below, are more common.

Throttling control can be implemented by using a variable speed driver. The speed of the driver, such as a steam turbine or gas (methane) engine, is continuously controlled in response to the system pressure. This method normally is limited to reciprocating and centrifugal compressors.

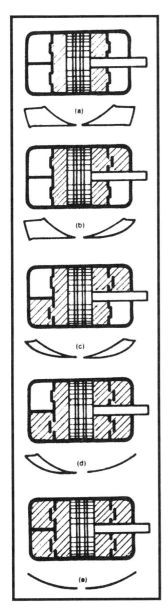

3.3.2.4 Type D—Modified Throttle Controls

A modified form of the throttling control described above frequently is applied to rotary compressors, both positive displacement and dynamic. The compressor output is varied continuously in response to changing system pressure over a region from a practical minimum capacity of 50-70 to a maximum of 100 percent capacity. At a preset high pressure, the compressor unloads and delivers no output, but continues to run at a reduced electrical load. See Figure 3-15, Diagram D, for pressure/output diagram. The throttling control for rotary screw compressors frequently is a valve in the air inlet which restricts air flow in response to compressor output pressure. The compressor output is reduced as pressure rises. By closing the inlet valve completely as a preset high pressure is reached, the compressor is unloaded and airflow reduced to zero.

Centrifugal compressors, by design, tend to decrease their output as the system pressure increases, such increase generally associated with a dropping demand for air. If this process is permitted to develop to dis-

Figure 3-16 Five Step Control Using Clearance Pockets and Inlet Valve Depression
Ingersoll-Rand Co.

charge pressures high enough to induce surge, compressors may be damaged. Such compressors normally are equipped with a by-pass set to operate somewhat below the surge point. This by-pass diverts output from the compressor discharge to atmosphere, thus reducing the amount of air available for distribution and matching output with demand. Other centrifugal compressor controls include inlet throttle valves and inlet guide vanes which permit capacity modulation over a range of about 70 to 100 percent of maximum.

3.3.2.5 Control of Centrifugal Compressors

Variable speed motor drives are available to control both output and discharge pressure. This is a more costly control system, but it may be most economic as it is the most efficient (See Section 5.2 below).

Adjustable inlet guide vanes are the most frequently applied control. While not as energy efficient, they make possible excellent part load performance and a wide control range. They can deal with significant changes in demand pressure as a function of flow. They perform well for medium and high ratios of compression.

Another control system, especially useful when demand pressure is more constant, makes use of adjustable diffuser vanes. Again this system works well at medium through very high compression ratios. It performs well over a wide output demand range.

A combination of variable speed motor drive with adjustable diffuser vanes is sometimes best, and combining inlet guide vanes with adjustable diffuser vanes is recommended where compression ratios are very high and only medium volume control range is needed in the presence of widely fluctuating system pressure. Simultaneous control of inlet guide vanes and adjustable diffuser vanes is energy efficient and offers excellent control over a wide range of conditions.

There are many good options for controlling centrifugal

compressors, and these installations normally are large, using a good deal of power. It is therefore advisable to study carefully all of the options and work with the compressor supplier to settle upon the best choice for the installation in question. Every application is different.

3.3.2.6 Surge Control for Centrifugal Compressors

Surge control devices are divided into three categories: surge detection, surge control, and surge anticipation. Surge detectors determine when a surge has occurred, surge controllers react to prevent further surges, and surge anticipators prevent machines from reaching surge.

Machines of different designs and different ages have different surge tolerances. Many centrifugal compressors are not hurt by limited surge. When surge damages a compressor it is usually due to the rapid pressure fluctuations against the rotating shaft, which can damage the thrust bearing if it is old or poorly engineered. Surge also results in recompressing gas which was already compressed, which can heat the gas to unusually high temperatures. But in one experiment a five hundred cfm compressor was attached to a ten-foot closed pipe, so surge began immediately upon startup. After half an hour the system reached equilibrium temperature, and even after seventy-two hours of continuous operation the compressor was unhurt. So the danger of surge depends upon the specific design of the compressor being considered and the manufacturer must be consulted.

Surge anticipators are the most functional and most often used; they combine surge anticipation with the steps used by surge controllers to prevent further surge. Compressor performance is characterized so the parameters just before surge are recognizable. For constant speed compressors the parameter used is generally air flow through the compressor. The compressor will surge at the same flow rate each time, so, as the flow rate approaches surge, the release valves are opened. However, flows are hard to measure. Differential pressures across the machine are usually measured, and the pressure ratio gives a good surge indication. This technique can be used on any dynamic compressor.

The motor current (amperage) of the compressor also is influenced by pressure and temperature. Using motor amperage to anticipate surge is particularly effective for low-pressure systems.

These controls can be made built around either simple analog or computer controllers. The paradox of surge anticipation is its attempt to control two effects at once: discharge pressure or mass flow control, and surge control. At times these priorities are in conflict, and tuning the surge controller is important. In most cases protecting the compressor is the most important priority. In other cases, for example in chemical plants where process interruptions are highly expensive, continuous flow is more important.

Surge controls are used almost universally in non-industrial applications. The added cost of surge control is not great, so their use is increasing in industry.

3.3.2.7 Multiple Compressor System Controls

Many compressed air systems call for more than one compressor operating simultaneously. How these compressors are controlled and operated, loaded and taken on line and off in response to demand, will be crucial to energy efficiency. Older control methods used sequencing to establish operating compressor(s) selection and more modern systems use a centralized control.

Sequenced multiple compressor systems are widely used. In these systems it usually is desirable to program all but one compressor to operate at full load so as to maximize efficiency while, at most, one compressor is cycled for modulated for output control. These multiple compressor systems can be classified by their control approach as pressure or pressure and time systems.

The simplest sequencing systems are pressure controlled only, sequentially loading and unloading different compressors in a manner similar to Figure 3-17, Part A. The five compressors are of the same capacity and each has Type A controls. The overall pressure control range PH-PL is three to four times that of each individual compressor preset control range. Narrow pressure bands will result in undesirably frequent cycling. Coincidence controls normally are not needed to prevent simultaneous starting of motors.

The pressure and time control systems incorporate one or several time delays in the start/stop cycling of the various compressors. Recycle timers are used to reduce the frequency of starting. Some set a minimum on or minimum off time, i.e., suppressing stop or start signals from the pressure switches. More complex controls use a combination of two pressure switches and timers to reduce the system pressure band. Compressors are started or stopped only when the preset pressure limits are exceeded for a fixed time as shown in Figure 3-17, Part B. The control determines which compressor to turn on or off. The delay time at the pressure extreme normally is a variable, with optimum value determined by plant load variation. Some more complex systems vary the time delay in proportion to the deviation of pressure from the desired limits. The system then responds quickly to large deviations and very slowly to small deviations from the preset pressure limits.

For both pressure-only and pressure-and-time control systems, the starting sequence can be adjustable to equalize the running time of the different units. High capacity and high efficiency compressors may be set up to carry the base load under normal operating conditions, but equal size units should be rotated in the sequencing. The ordering of compressor sequencing can be manually controlled with transfer switches or automatically controlled with a programmable sequencer.

The more modern approach makes use of a solid state master system controller which sends signals to the individual controller on each compressor. If the master fails, the individual controls on each compressor will allow the system to continue running.

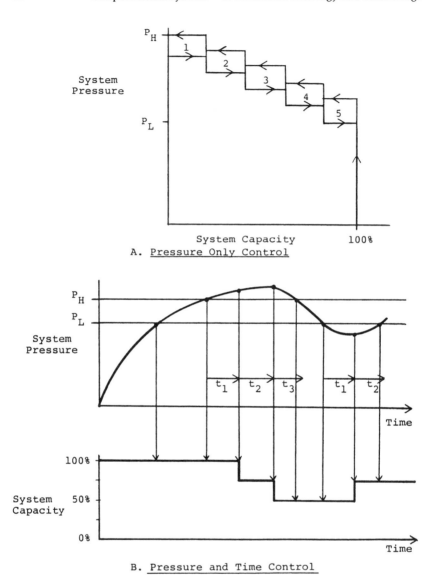

A. <u>Pressure Only Control</u>

B. <u>Pressure and Time Control</u>

Figure 3-17 Multiple Compressor Control Systems

Many control and decision making system designs for multiple compressor central controllers are available. Most are dependent on PLCs (programmable logic controllers, or computer chips), and the best have easy manual overrides so, in the unlikely case of chip failure, the system still functions.

The simplest systems have all compressors working in parallel and open or close the inlets the same amount on each. As pressure demand rises, each compressor gets loaded at the same rate and does the same amount of work. Similarly, the compressors are backed out in parallel as demand falls.

Slightly more complicated systems, best for maintaining pressure in the system but sometimes less perfect for energy conservation, have a transducer connecting the pressure system to the PLC. The PLC brings the next compressor on line or takes one off line when the transducer output rises or falls past preset thresholds. This system works well for two to eight compressors, and it functions to keep the pressure in the total system constant. Without transducers, each compressor must have its own pressure switch and the system becomes more complicated and more susceptible to failure.

Whether each compressor comes on fully loaded or in gradual steps depends on the controls for that specific compressor. Centrifugal compressors load up gradually, while vane compressors come on fully loaded.

In an alternate configuration, the compressors are set up identically and operate in sequence just as in the previously discussed electrically controlled system. The first machine begins as the master machine. Its own controls load it more or less as demand rises and falls. When it reaches full load and remains there for a set amount of time, it starts a second machine, which then becomes the master machine. The system reacts to reducing demand similarly, unloading the last machine but leaving it armed for a set amount of time, and then shutting it off and beginning to unload the next in line. This system is designed for maximum energy conservation.

Similarly, the system can be designed with a constant master compressor which controls pressure, and a series of permanent slave compressors which cycle through being unloaded, minimally loaded, and fully loaded, depending on demand.

New installations generally are equipped with multiple controls, but older installations usually are not upgrading quickly. Multiple control systems are easiest to install with new installations, where the control systems for the individual compressors send out more consistent signals. In old systems, especially those which have been expanded over time, the inaccuracies in each compressor's transducer are exaggerated and accurate control is more difficult.

Other types of multiple compressor controls include duplexes and triplexes. In a duplex control, two compressors are alternated between lead and lag. In this system there is always a demonstrably working backup compressor, and equal wear is put on each. In a triplex system the three compressors are rotated through lead, lag, and backup.

Central controllers also are available for situations in which the multiple compressors are not all the same location. These are much more difficult to control, and sometimes multiple pressure transducers are needed.

3.4 Aftercoolers

Some air systems can and do distribute and use uncooled air, just as it leaves the final compression stage. Certain forges, foundries and processes can use this air, which is both hot and wet. However, such installations are rare; almost all systems are equipped with aftercoolers. These both lower the temperature and remove water from the air.

An aftercooler is always recommended for instrument and general plant air. Among the criteria applied to the selection of aftercoolers are: required discharge temperature, required dryness, air inlet temperature and humidity, coolant conditions at aftercooler, temperature at using point, whether or not the aftercooler is followed by a dryer, availability and cost of cooling water, and cost of (blowing) cooling air. Normally, aftercoolers are designed for the worst combination of conditions. When a dryer follows the aftercooler, the dryer will have a maximum recommended inlet temperature and this specification must be met.

Aftercoolers will produce a considerable amount of condensate most of the time. Therefore, they should be equipped with separators and traps as has been noted elsewhere. Frequently they have built-in separators and sometimes they come as part of a compressor design package. After the above basic criteria have been met, the user always has the option of oversizing the aftercooler for both reduced pressure drop and for further temperature reduction. Oversized aftercoolers frequently are a good investment when they are followed by dryers because of possible reduced energy consumption of the total system and the reduced fouling problems during use.

Ease of aftercooler maintenance should be considered. Some tube and shell water cooled aftercoolers can be readily disassembled for regular cleaning.

The consumption of water can be reduced significantly if aftercooler water flow controls are installed. These almost always are good investments and, just as in the case of compressors, the coolant flow rate control valving should be located on the water discharge side of aftercoolers.

3.5 Heat Recovery

Most of the prime mover energy applied to compressing and conditioning air is lost as heat carried off from the compressor cooling system, the intercooler, and the aftercooler. When this heat can be recovered and used, then significant cost savings are effected.

Many compressor companies make available heat recovery packages which will permit the directing of some of this heat into the heating, ventilating, and air conditioning (HVAC) systems of nearby buildings. The engineering of such systems always should include HVAC considerations and HVAC engineers. The economic and energy tradeoffs of these systems will be discussed in greater detail in Section 4; but suffice to say that, for all compressor installations of 50 or more horsepower, heat recovery should be investigated.

3.6 Receivers

Receivers perform several functions in compressed air systems. First, they provide a larger system capacity, which increases the cycle time of compressor control systems. This makes less difficult the elimination of unstable and overcorrecting control cycles.

The receiver also dampens pulsation from reciprocating compressors, acts as a reservoir to prevent excessive temporary pressure drop during sudden short-term demand, and can be used to smooth air flow through dryers, separators and other air conditioning equipment. Because the air entering the receiver is reduced in velocity and cooled, some of the moisture may condense and fall to the bottom of the receiver where it can be removed by a valve or, preferably, a trap. Such a receiver can reduce further the amount of moisture which must be removed by a subsequent drying stage. The receiver always is equipped with a pressure relief valve.

The receiver size in cubic feet normally should be at least 12 seconds of compressor capacity in small units,to not less than 8 seconds of compressor capacity for large units.

Sometimes smaller compressors are mounted upon receivers and the receivers serve as a basic mounting frame for the assembly of the compressors and their accessories. Under these and some other circumstances, the only pressure relief valve frequently is located upon the receiver. When this mounting procedure is employed, care must be taken to assure that no other valve is between the compressor and the receiver, and that support members or frames are not welded into receiver tank seam welds. Inspect carefully and refuse to take delivery of unsafe equipment.

3.7 Filters and Separators

Filters and separators are used to remove contaminants, especially particulate matter and oil and water, from the air. Different types are used for different requirements and locations within the system.

3.7.1 Inlet Filters

It is desirable that particulate matter be prevented from entering the compressor. Air inlet filtration will reduce compressor maintenance and possible compressor damage due to ingestion of medium and large particles. Also, proper filtration will prevent contaminant buildup on the impellers of dynamic compressors resulting in depression of surge margin and requiring preventive maintenance.

Three types of filters normally are used and in increasing order of efficiency and they are: viscous impingement, oil bath, and dry.

The first depends upon a base of multiple layers of screen or fibrous material coated with a liquid, frequently oil, which causes particulate matter to adhere to the filter base. Maintenance includes cleaning the base and replacing the oil coating. Oil bath filters draw unfiltered air through an oil reservoir and then through a screen mesh or other elements which clean out the oil and the dust particles as they adhere to the oiled element surface which is continually washed by oil circulation. The dry types simply strain contaminants from the air through the use of densely spaced materials which simply block particulate passage.

The first two types are not suitable for use with oil free compressors, but may be less costly in service if the cost of cleaning is less than that of filter replacement. All can function as filter/silencers, and also are offered as in-line air filters by some manufacturers. When in-line air filters are used, then the air intake itself should be equipped with a screen and weather hood. It is possible that two stages of filtration may be indicated for some compressor types, particularly non-lubricated designs, because of the sensitivity of these compressors to the buildup of contaminants on impellers, screws, and cylinder walls.

In the selection of filters, consideration must be given to the pressure drop which is likely to occur across them. This should be balanced against the maintenance costs of both the filter and the compressor. Too great a pressure drop across the inlet filter reduces both the compressor mass flow capacity and its part load operating efficiency. A drop of 12 inches H_2O reduces capacity about 3 percent. The clean element pressure drop across most inlet filters normally will vary from 1/4 inch to 10 inches of water depending upon the type selected and the filter size. Consideration should be given to increasing the size of the filter in order to assure maximum compressor capacity.

3.7.2 Compressed Air Filters

Air is filtered after compression in order to remove oil, water, and any particulates produced by the compression process. The nature of this filtration equipment will be determined by the system air quality specification. It is not economical to provide contaminant removal to an extent greater than that required by the using system. Once this specification has been established, manufacturers can offer a selection of suitable contaminant removal equipment. This may include particulate removal filters, coalescing (coalescing small particles and hydrocarbon mists into larger droplets for separation) filters, adsorption filters or even activated charcoal filters for removing odors and hydrocarbons. It is to be noted that the adsorption filters can remove vapors with less pressure drop than the other types. Some manufacturers combine particulate and coalescing filters into one unit.

Filters should always be equipped with drains and should be installed at the place in the system with the highest pressure. Their specific location is dependent upon the intended service, the type of filter and the type of dryer. Filters should be selected carefully as they may be the highest pressure drop component in the system. Careful sizing and selection can save 2 to 6 psid, which is 2 to 6 percent of the pneumatic energy available.

When dryers are included, the filtration may be more critical. A desiccant type dryer normally will require a high quality coalescing filter installed upstream in order that the adsorption material not be contaminated. A 3 to 5 micron particulate filter normally will be installed downstream from such a dryer to prevent adsorption materials from being carried into the distribution system. Some refrigeration dryers require coalescing filters upstream in order to keep the surfaces of the heat exchangers free of varnish. If a further reduction of oil carry-over is required downstream of the dryer, then another coalescing filter should be installed after the dryer.

Often filters are combined with separators in a single unit.

3.7.3 Separators

Separators should be installed in the system at any point where it is desirable to remove entrained liquids. Depending upon inlet air quality, this might be before the first stage of compression or immediately following a compressor intercooler. Installation of separators after air compression will also reduce the content of oil and water in the compressed air.

There are several types of separators, one being a gravity type. This one is not popular because normally it consists of a large volume to permit gas velocity to be lowered so that particles will settle out. This large volume requires space and is costly.

The most popular types are those which separate by rapidly changing the direction of the air stream so that the particles will be centrifugally separated from the air. These are impingement, centrifugal, and cyclone types. Normally these have a low pressure drop, although some are better than others in this regard. They are designed for a specific flow rate since the flow rate determines the velocity and hence, the centrifugal force; they should not be oversized. If the flow rate is expected to be highly variable, then it may be desirable to connect several separators in series, each designed for a different band of air flow rates.

All separators should be equipped with traps or drains.

3.7.4 Air Line Filters

Frequently, additional filtration is added to accommodate specific air usage. For example, combined filters/ lubricators usually are installed immediately preceding air tools. These remove any particulate matter which may have been generated in the distribution system and would be damaging to the tool. The lubricator portion will provide lubricant for the tool.

3.8 Traps and Drains

Traps and drains collect liquids which have been removed by cooling, separation, filtering, or condensation and automatically will release this condensate, while minimizing the discharge of any compressed air. Traps should be located at any point where moisture, for any of the above or other reasons might be collected. This will include separators, filters, coolers, receivers, and dryers. They also should be located at any low point in a distribution line especially if that line is passing through a cold area.

A properly installed trap will be bypassed by a manual valve, and will be preceded by a strainer with a blow-down valve, a pipe union, and a cut-off valve. These various accessories to the trap installations permit servicing the trap and blowing debris from the strainer. The strainer prevents particulate matter from fouling the trap.

3.9 Dryers

The amount of moisture which can be absorbed as a vapor and remain mixed with air is determined by the temperature and volume of the mix. Pressure is not a factor. The implication of this is important for the operation of compressed air systems.

Even in dry climates, air supplied to the inlet of air compressors almost always will contain some water vapor. The quantitative measure of this vapor is specific humidity. Relative humidity is a measure of the amount of vapor in the air compared to the absolutely maximum amount that the air can absorb at that specific temperature. Since air at higher temperatures can hold in absorption more water, heating the air, without changing the moisture content, will lower the relative humidity and vice versa. When air is cooled to the point that the relative humidity reaches 100 percent and then is cooled further so that some of the moisture barely begins to liquefy, then the air is said to have been cooled to its dew point.

When air is compressed, the volume is reduced and hence a great deal of moisture that had been absorbed in a large volume of air finds itself compressed into a much smaller space. Since raising the pressure does not increase the amount of moisture which can be absorbed in that volume, then compression has the effect of raising the specific humidity. However, since most compression processes are nearly adiabatic, the temperature of the air rises, permitting more water to be retained temporarily in absorption by the now higher pressure but much warmer air. The problems come later in the system when the air is cooled.

Some of this water will be removed by inter-cooling, aftercooling, separation and filtration as the air drops in temperature throughout these post-compression conditioning phases. If no further drying is applied to the air, it will then enter the distribution system at a relative humidity of 100 percent and usually at a temperature higher than many portions of the distribution system, tools and pneumatic devices. The result is system wide condensation, pipe and valve corrosion, higher maintenance costs of tools and pneumatic devices, contamination of paint and finishing processes, and other undesirable effects of having condensing moisture accompany every usage of the compressed air. As a result, there has been a growing tendency to dry the air,and for this purpose several different types of dryers are offered by industry.

The oldest, lowest initial cost, and most energy efficient is the deliquescent dryer, which simply absorbs moisture into deliquescent material which must be replaced periodically. Replacing this material is labor intensive and may require either system shutdown or temporary operation without drying while materials are being added. Typically, these can be designed to drop the dew point by 12 to 20°F when salt is the deliquescent material. Using potassium carbonate permits drying to a drop of about 36°F.

A much more common class of dryers use refrigeration, and these can lower the dew point to about 35°F.

Desiccant dryers can provide the lowest dew points of any type, to –40°F. These are initially the most costly dryers, and the least energy efficient. They adsorb the water vapor into adsorbing materials, which are then regenerated either by electric heating, air purging, or both. Some of the more modern units now also offer vacuum purged regeneration and these use less energy.

In most cases, the type of dryer needed is determined by the pressure dew point required in the system.

3.10 Distribution System

The distribution system is that part of a compressed air installation which takes the air from the compression and air conditioning equipment and distributes it to its various end users. It is one of the least understood and most important parts of the total system, since it is the place where so much energy is lost and so much maintenance is incurred.

Upon leaving the air conditioning portion of the system, the air passes through a riser which takes it up to the header or headers. Compressed air distribution systems normally are overhead for routing convenience and availability for service and draining. The headers are the main distribution lines and carry the air to the subheaders or branch lines which then serve all of the various drop points. All lines should be pitched to low points where drip legs, with traps or drains, are installed. At or near each use will be a drop point or drop station which is a vertical line bringing the supply down to the tool or work level. At the end of the drop line will be whatever fittings are necessary for connecting to the end use. These lines also should be equipped with traps or drains at their low points. In some cases a stationary tool will be hard connected to the piping through suitable valving and sometimes strainers, filters, separators, etc. In other cases, the drop will be equipped with one or more quick disconnect fittings so that hoses can be snapped into place. The hoses then, in turn, serve portable pneumatic devices such as portable hand tools. Each such drop then may be equipped with a combination filter/ lubricator assembly or a filter/regulator/lubricator assembly.

Sometimes the main headers are configured in loops providing multiple air passage to the more distant portions of the system. These do tend to offer a more uniform pressure at these distant points. It is frequently recommended that the headers be significantly oversized, not only to reduce pressure drop but also to provide a better system time constant (capacity) for more economic control of compressor operation. Some engineers argue that purchase of capacity through oversizing headers is less costly than an equivalent capacity in the form of a receiver.

The subheaders or branch lines should be sized for minimum pressure loss as should be the drops and hoses.

Some installers recently have been using plastic pipe in compressed air systems as it is lighter and hence much less costly to produce and install. However, there is a great deal of uncertainty in the industry about the wisdom of this for a number of reasons. The following shortcomings have been reported:

1.   Polyvinyl chloride types give off toxic vapors when subjected to fire. Furthermore, they will support combustion.

2.   There is a good deal of thermal expansion in plastic pipe and special design considerations must be taken to avoid excessive stress.

3.   Some lubricants attack plastic pipe.

4.   Polyvinyl chloride pipe dries out and flakes in the presence of compressed air.

5.   Plastic pipes tend to creep under stress.

6.   In a general plant fire, the high temperatures might very quickly destroy the pipe and have it release compressed air, thus further aggravating the fire.

Thermoplastic pipe is not recommended for exposed, above-ground usage for transporting compressed air.

If plastic pipe is to be used anywhere in a compressed air system, it might be permissible for inlet lines. The risers, headers, subheaders, and branch lines should be restricted to welded, threaded, or flanged steel pipe.

The matter of maintaining a constant system pressure is important since so many pneumatic devices suffer in efficiency when they are subjected to pressures which are under specification. On the other hand, if, in order to maintain such pressures, it is necessary to overcompress and then reduce pressure, this in turn represents a significant added operating cost. Therefore, a distribution system which has a volumetric capacity adequate to minimize pressure drop will provide long term economy. System layouts sometimes can be improved by having more than one location for receivers and compressors. However, this does complicate compressor control.

Distribution systems frequently are equipped with certain specialized types of valving. Invariably, there will be pressure relief valves protecting all of the receivers. In addition, relief valves are usually located in the systems near the compressors, particularly positive displacement designs. However, the location of such a safety valve must be in the discharge line or receiver at a point upstream of any other cutoff valve. All compressors should be equipped with check valves located in the discharge line and preferably located near the compressor discharge and in a horizontal position, upstream of the aftercooler and moisture separator.

Some systems, in order to maintain more nearly constant pressure, will be equipped with reducing valves. This permits the storage of air at a pressure higher than that at which it is used and consequently increases cycle time for the compressor or compressors which are operating automatically in response to pressure. The problem with this type of control is that the air is being compressed to a pressure much higher than is necessary for use and the resulting loss of energy is essentially proportional to this percentage of overpressure. Therefore, this practice is not recommended and better control methods almost always will be economic.

A wide variety of manual valve types and connectors are available, and the choice of these should be directed toward avoiding excessive pressure drops and leakage in the system. This will be discussed in greater detail in subsequent sections of this book.

3.11 Pressure Regulators

Pressure regulators are employed to supply air to pneumatic equipment at pressures lower than the supply pressure. Various types of regulators are available to supply a wide variety of applications including:

- Low flow, precision regulated instrument air
- High flow, reduced pressure pneumatic tool supply
- High flow, pressure regulated air cylinder supply
- Pressure reduction for cleaning and blowing

Regulators should be selected to serve a specific application. The proper regulator selection is based upon supply and secondary pressures, air flow rate during equipment usage, type of equipment supplied and pressure accuracy required.

Regulators can be properly used to supply small volumes of

air at pressures lower than the system supply. If large quantities of low pressure air are required, a separate low pressure system should be considered as it may offer long term economic benefits.

The reduction in pressure achieved by a regulator results in a reduction in the energy content of the air, i.e., air compressed to 100 psi and regulated to 60 psi at the tool site results in approximately a 40 percent energy waste. An inadequate distribution system should not be offset by increasing the compressor pressure setting and installing regulators at a large number of tool sites. Better solutions are offered in Section 4.2.3.

Some pneumatic devices, such as air cylinders, require a specific volume of air to operate and can adequately perform their task at reduced pressures with a net savings in air consumption. Typically, the operating pressure for air cylinders is determined by the task to be performed. Frequently the retract cycle is a non-working motion which can be accomplished at a lower pressure through the use of separate regulators On the extend and retract cycle. This type of installation can effect considerable energy savings at low cost.

4.0 Energy Saving Considerations In Compressed Air Systems

The rapid increase in energy costs over the last decade has changed the way in which energy consuming equipment is specified and applied. In the past, energy was cheap, and the first cost of equipment was the major expense; therefore, increasing the capacity of a compressed air system was simply a matter of adding an additional compressor. At current utility rates of up to 20¢ per kilowatt-hour, the electrical costs for operating an air compressor have become the major component in the total cost of compressed air over the lifetime of the compressor. The first year's utility bill for operating an air compressor can exceed the purchase price of the unit. For example, a 100 kilowatt air compressor operating 6,000 hours per year at a cost of 6¢ per kilowatthour would use $36,000 worth of electricity. In this context, energy consumption must be a primary consideration in any compressed air system design.

It is important to consider specific areas of compressed air systems in which energy can be conserved. Section 4.1 discusses each major energy consuming component of the system individually in terms of selection, sizing, and operation for reduced energy consumption and minimum life cycle cost. Section 4.2 suggests energy conserving practices which can be applied to new system design and system modification.

4.1 Energy Characteristics of Major System Components
Most components of any system can contribute to energy

71

loss in one way or another, either directly or indirectly. The most conspicuous energy users in an air system are the compressor, the conditioning equipment and the distribution system.

4.1.1 Compressor Efficiency

The specific efficiency of an air compressor under full load conditions is measured in units of bhp/100 icfm when compressing to 100 psig. Compressors of the same type and size will have similar specific efficiencies, with some variations between manufacturers and design details. Typical values of specific efficiency versus motor horsepower capability from 10 to 10,000 are shown in Figures 4-1 and 4-2. These are only typical values with actual specifications of manufactured models falling on either side of the indicated values. The user should obtain current data on several compressor models from various manufacturers to make valid comparisons. Care should be taken to obtain the actual output motor bhp and not just the faceplate rating, since some compressor motors are sized to operate at a service factors greater than nominal full power. Figures 4-1 and 4-2 should serve only as a general guideline; verified performance information for the actual compressors to be considered should be used.

This is only part of the story, however. There are few installations in which compressors operate at full load all of the time. Most factory operations are such that the loads are highly variable, and most of the time, most compressors do not operate at full load, because they normally are sized to handle peak requirements. Reduced load performance is provided in several ways which will be discussed in Section 4.1.3. Some of these will require that the compressor operate at no load or idle condition some of the time. Others will provide for modulation of the compressor output by one of several different means. The performance at part load is determined by a number of factors, such as compressor type and control methodology used, as will be shown. Compressors should be compared on a basis of the full operating profile of expected air use. Any other comparison criterion, such as efficiency

Figure 4-1 Relative Full Load Power Required of Typical Lubricated Compressors, 100 psig, at Sea Level

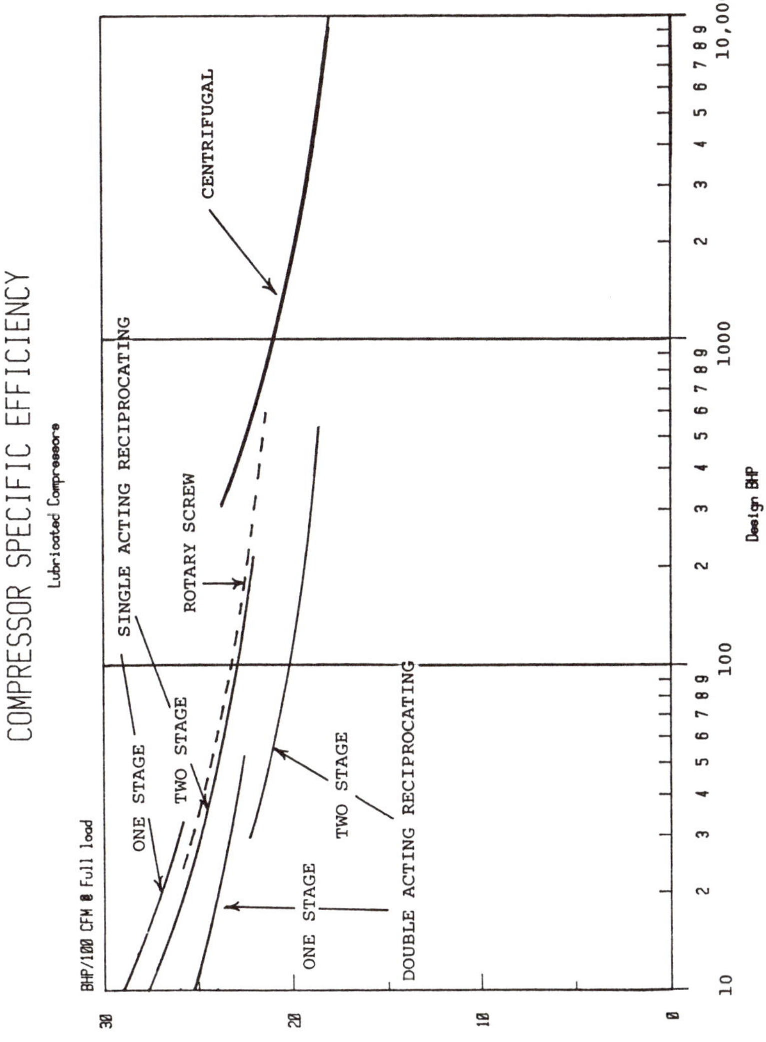

COMPRESSOR SPECIFIC EFFICIENCY

Lubricated Compressors

Figure 4-2   Relative Full Load Power Required of Typical Oil-Free Compressors, 100 psig, at Sea Level

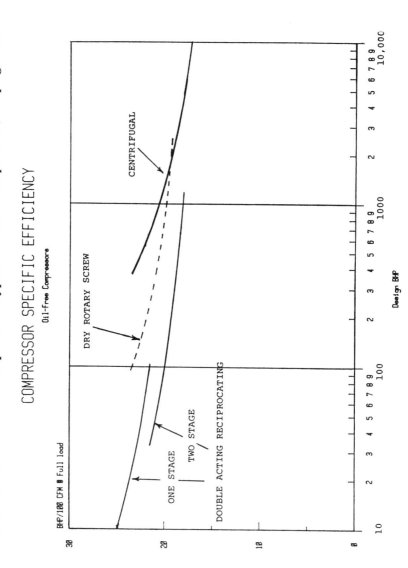

at full load, will be misleading and could result in excessive lifetime costs.

For example, it is not infrequent that compressors are started one half hour to one hour before production is up to full tempo in most factories. This is to provide air for instruments, for checking out tools, for blowing down traps and for other housekeeping functions preparatory to a full production operation. During these times the system is providing essentially only that air necessary to feed the leaks, which might be 10 to 25 percent of full load capacity.

Then, as the day progresses, the pressure usually is kept up through coffee breaks, lunch period, and temporary line shutdowns. All of these cause extended periods of compressor operation at reduced load. When shifts, lunch and other breaks are staggered (not all departments necessarily break at the same time), then there might be a period of 1 to 2 hours in the middle of the day and at shift changes when only about half of the peak compressor capacity is needed.

As the workday nears the end, some operations shut down before others. In some factories workers are permitted to depart as soon as their particular line has achieved a quota. One team may leave, then another, until finally when the last bit of production has been complete the air system can be shut down. It is not unusual for a single shift operation to keep the compressor operating for a total of 10-1/2 to 11 hours, during which time it will have been operating at full capacity for only 20 to 30 percent of the time.

It is obvious then that part load performance is of immense importance to overall energy consumption. Figure 4-3 shows energy efficiency comparative information for one typical profile of factory operations. This presentation was developed using:

1.   The specific operating efficiency of each compressor type presented in Figures 4.1 and 4.2.

2.   The part load efficiency factors for each control type presented in Table 4.1.

3.   Typical motor efficiencies at reduced loads from Figure 4.6.

These calculated results are typical values only and are not
currently guaranteed by any manufacturer. Here it can be seen
that the efficiency differences between compressor types can be
accentuated and the energy and cost implications of these differences become significant.

It is important then, when selecting a compressor, that the
operating profile of the factory be investigated and that all candidate compressors be analyzed with expected true operating
conditions in mind.

The one exception might be multiple compressor installations.
It is not unusual, in these situations, to have some of the compressors operate continuously at full load, programmed for the
minimum factory demand, and other compressors in a variable
output mode. A reasonable control program might include using
controllable, double acting, two-stage compressors until demand
reaches a level suitable for a centrifugal unit. The centrifugal
compressor would then be used to handle the base load with the
other compressors modulated to meet the variable demand beyond
that supplied by the centrifugal unit. There are many scenarios
that could be found to be workable for any one installation. Each
should be evaluated for lifetime cost.

Further insight into the analysis of full and partial load
profile performance, can be found in Sections 4.1.3 and 5.2.1.

4.1.2 Driver Efficiency
The vast majority of plant air systems are powered by electric
motors. Some special circumstances may favor alternate drivers,
as indicated in Section 3.2, but almost all of the energy used to
compress air passes first through electric motors.

Electric motor efficiency typically ranges from 80 to 96

Figure 4-3 Relative Power Required by Single Compressor Supplying Variable Load, 100 psig, Sea Level

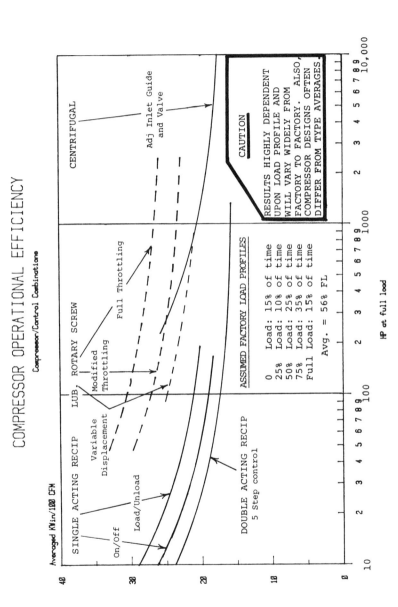

COMPRESSOR OPERATIONAL EFFICIENCY

Compressor/Control Combinations

Averaged KWin/100 CFM

CENTRIFUGAL

Adj Inlet Guide
and Valve

CAUTION

RESULTS HIGHLY DEPENDENT
UPON LOAD PROFILE AND
WILL VARY WIDELY FROM
FACTORY TO FACTORY. ALSO,
COMPRESSOR DESIGNS OFTEN
DIFFER FROM TYPE AVERAGES.

LUB. ROTARY SCREW

Full Throttling

Modified
Throttling

SINGLE ACTING RECIP

Variable
Displacement

On/Off

Load/Unload

DOUBLE ACTING RECIP
5 Step control

ASSUMED FACTORY LOAD PROFILES

0	Load:	15% of time
25%	Load:	10% of time
50%	Load:	25% of time
75%	Load:	35% of time
Full	Load:	15% of time

Avg. = 56% FL

HP at full load

10   100   1000   10,000

percent, with recent high efficiency designs now leading an upward trend. However, it is interesting that older designs (from 1950-1960) typically had efficiency as high as or higher than those of more recent designs. Evidently low energy costs, beginning about 1960, induced markets for less expensive motors, a situation now evidently changing.

Figure 4-4 shows the variation of efficiency vs. horsepower of one manufacturer's lines of motors. Both standard and high efficiency squirrel cage induction, 1800 rpm Class B, open frame motors are shown. The variation in motor efficiencies among different manufacturers is widespread, sometimes greater than the difference between high and standard efficiency models, as can be seen in Figure 4-5.

Just as in the case of compressors, it is important to consider the efficiency of motors at part load as well as at full load. Also, the nameplate ratings of motors do not necessarily indicate exactly how they might best be matched to compressor requirements. Some compressor manufacturers assume that most compressors will not operate continuously at full load, and hence they will specify smaller electric motors, planning that they be overloaded some of the time. Motor ratings are based on temperature rise (usually 40°C above ambient). As long as the motor is not continually operating at too high a temperature, no harm is done by overloading for short periods. In fact, a motor operating at high average loading and resulting higher temperatures will be more economic than one that is oversized. In practice most compressor motors have a service factor of 1.15. Hence, at full load, the motor is delivering 1.15 times its rated horsepower.

Therefore, in order to understand how much energy will be lost in the electric motor, the total operating profile of the system must be understood, i.e., the percentage of the time that the motor will be operating at various load conditions and also how the motor has-been matched to the compressor horsepower requirement. Figure 4-6 shows some efficiency/load curves for motors of several different manufacturers. The manufacturers should be

Figure 4-4   Average Motor Efficiency 1975, 1800 RPM Class B Open
Motors Standard, High Efficiency, and 1955 Designs

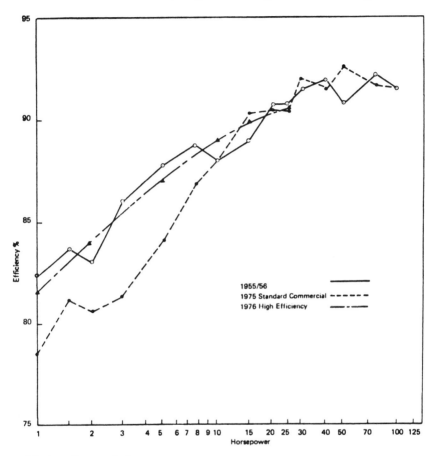

Varigas Research, Inc.

consulted and the efficiency curves on their motors evaluated at
the time the drive motors are selected. As can be seen from Figure
4-6, there is a great deal of variation in performance from motor
to motor, and how the motor is matched to the compressor can be
a large factor. It can be seen, as suggested earlier, that oversizing
the motor is more energy wasteful than undersizing it in most
cases.

Figure 4-5 Published Motor Efficiencies of Principal Manufacturers (Open, Drip-Proof, 1800 rpm, NEMA Design B)

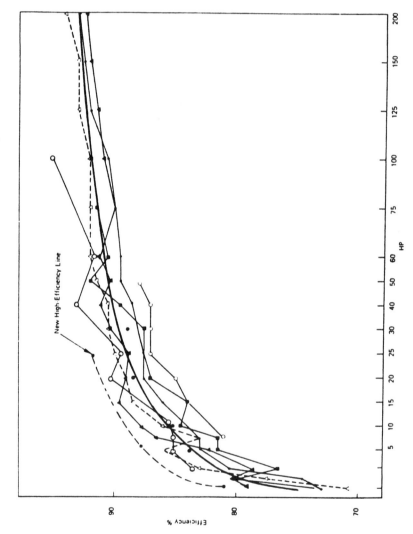

Figure 4-6   Efficiency vs Load Characteristics of Certain Standard Line Motors of Two Major Manufacturers (Horsepowers 5-125)

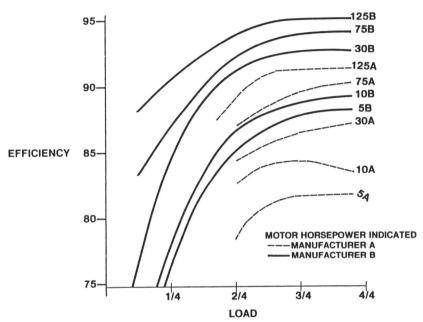

Varigas Research, Inc.

Once the operating profile of the system is known, motors of various performance characteristics can be compared, using the analytical program outlined in Section 5.2.1. In high usage situations (two or more shifts), the most costly, higher efficiency motors almost always can be cost effective. Sometimes it is true also of single shift operations, the final answer depending upon the cost of money and other inputs.

The way that the motor is controlled can effect additional reductions in power costs. Consider the four classes of losses in electric motors:

Core losses   Friction and windage losses
Winding (I^2R) losses   Stray load losses

The total of these for one typical motor is shown in Figure 4-7; between 50 and 60 percent of the total is accounted for by the winding (I^2R) losses and these accordingly provide 50 to 60 percent of the temperature rise of the motor. Therefore, if somehow this component could be reduced, then both efficiency and temperature would be improved.

Figure 4-7. Total motor losses vs. motor voltage, Westinghouse Standard design B, 40-hp, 460V, 60 Hz, 4-pole, squirrel cage induction motor. *(Courtesy Westinghouse Electric Corp.)*

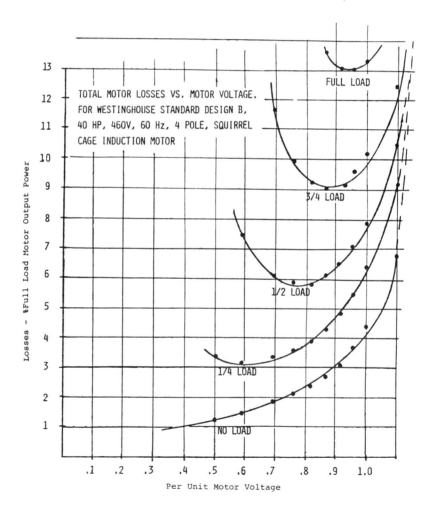

Variable voltage controllers accomplish exactly that. They use solid state devices, usually silicon controlled rectifiers (SCR's) to reduce the line voltage. Various designs sense the motor loading difference by measuring power factor, current flow, and motor current and slip. Such controllers are available for motor sizes from 2 to 500 horsepower, and frequently they incorporate other control functions such as reduced voltage starting, overload protection, undervoltage protection, and protection against phase loss. These controllers operate on the general principle that they sense the motor load, and reduce the rms voltage supplied to the motor terminals by clipping off portions of the ac wave form. The controllers that use power factor sense the motor load by comparing the time difference of the zero crossing points of the current and voltage to the motor, which is a measure of the motor power factor. When the power factor controller is applied to multiple horsepower, three-phase motors, serious instability problems frequently arise. The motor has difficulty recovering from low load conditions and can frequently stall. Other types of controls normally are used with multiple horsepower motors to avoid this problem.

The net power savings available at reduced loads for motors from 5-100+ horsepower are shown in Figure 4-8. Note that the power saved is measured as a percentage of full load power. If the saving is converted to kilowatts and then compared to the actual power being used, the savings are impressive under part load conditions. For example, assume a 20 horsepower motor which is 90 percent efficient at full load and 87 percent at one-half load. At full load the motor will draw:

$$20 \div 0.9 \times 0.7455 = 16.6 \text{ kW}$$

From Figure 4-7, at 50 percent load, the controller will be saving 3.9 percent of this power or:

$$0.039 \times 16.6 = 0.65 \text{ kW}$$

At 50 percent load the motor is drawing:
10 •. 0.87 × 0.7455 = 8.6 kW

The saving then is 0.65 ÷ 8.6 or 7.6% of input power

A controller which saves, on the average, 0.65 kilowatts each operating hour, year in and year out, can be cost justified in a number of applications.

Figure 4-8   Net Power Saved with scr Voltage Reduction Controller, Line Voltage: 483v

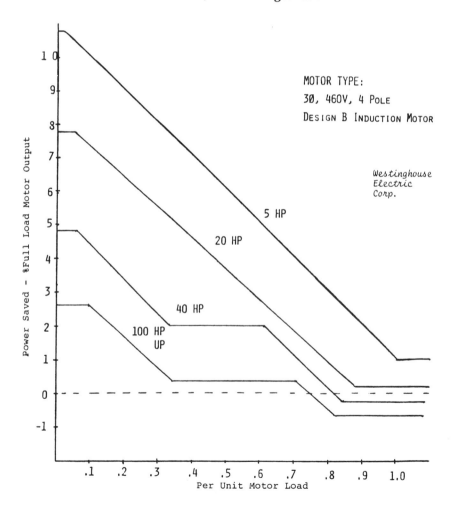

4.1.3 Compressor Control

Each combination of compressor type and control method presented in Section 3.3.2, has different energy characteristics at reduced loads. Table 4-1 indicates the power required to drive some compressors at reduced load. The modulating characteristics are indicated at the top of the table, and the percent full load for each step in the line "Modulating Range." The different types of compressors are noted in the first column. In the remaining columns, first are shown the percent of full load power drawn during each of the modulating steps, and then immediately beneath these numbers, in parentheses and italics, are shown the relative efficiencies. These relative efficiencies are the percent of full load efficiency, or the amount by which the power required per 100 scfm must be adjusted in order to compute the power required at the reduced loads.

For example, assume a double acting reciprocating compressor having a full load power requirement of 22.0 bhp at 100 scfm, and equipped with a Type B multi-step control. At 25 percent load, the power required will be 22 ÷. 0.78 = 28.2 bhp/ 100 scfm, at 50 percent it will be 22 ÷. 0.91 = 24.2 bhp/100 scfm, and at 75 percent load the power required will be 22 ÷ 0.96 = 22.9 bhp/100 scfm. Figure 4-9 provides some of the same information in graphic form for each of the four major types of controls. The solid lines indicate continuous throttling ranges while the dashed lines connect discrete operating points indicated by large dots. The lower curves indicate more favorable part load efficiencies.

It is again noted that these are only typical efficiency values with exact values for specific combinations of compressors and control systems falling on either side of the values in the table and figure.

Rotary screw compressor controls are typically a modified throttling type. Several mechanisms are available for performing the output control. Input throttling valves close off the inlet, causing a partial vacuum to develop at the inlet, both effectively increasing the compression ratio and reducing the volume of

Table 4-1 Compressor Efficiencies at Part Load as a Function of Compressor Type and Control Mechanism

Part Load Efficiencies (%FL bhp at Indicated Reduced Loads)

Control Type:	Type A - Two Step		Type B - Multi-Step		Type C - Throttling		Type D - Modified Throttling	
	Start/ Stop	Load/ Unloaded	Double Acting Unload	Unloaded/ Clearance Pockets	Discharge Bypass	Variable Speed	Inlet Valving (Same w/Un-loading)	Modified Thrott-ling Inlet Guide Vane Adjustment
Modulating Range (%FL Capacity)	0, 100	0, 100	0, 50, 100	0, 25, 50,	0-100	20-100	0, 40-100	0, 70-100
Reciprocating Compressor Single Acting	0, 100 (100%)	10, 100 (0, 100%)[2]				19-100 (105-100%)		
Double Acting	0, 100 (100%)	10, 100 (0, 100%)	0, 55, 100 (0, 91, 100%)	10, 32, 55, 78, 100 (0, 78, 91, 96, 100%)		15-100 (130-100%)		
Rotary Compressor Helical Screw, Lobe, Rotary, Vane	0, 100 (100%)	20, 100[1] (0, 100%)				For 50-100% load 60-100 (85-100%)	20, 80-100[1] (0, 50-100%)	

		For 30-100% load 50-100 (60-100%)		For 70-100% load 16, 75-100 (0, 93-100%) For 50-100% load 16, 62-96 (0, 81-96%)	11, 71-100 (0, 98-100%)
Centrifugal Compressor		100,100 (0,100%)			

Notes:

1. ( ) denotes relative part load efficiency (%load/%FL capacity).
2. Assume sump bled down when unloaded. If sump not bled down, 70% FLbhp when unloaded.

Figure 4-9   Percent Full Load Power Required for Reduced Output

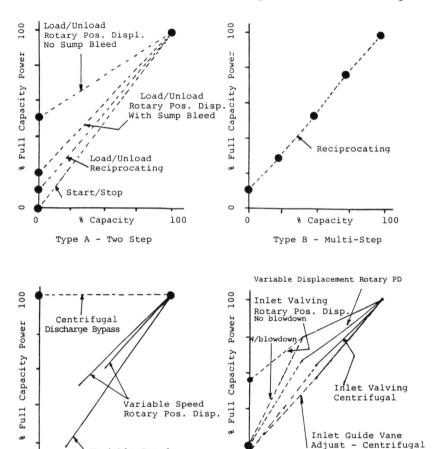

compressed air delivered to the discharge. Inlet throttling can reduce output volume to 40-60 percent of capacity with a corresponding power requirement of 70-85 percent full load (FL). If the inlet valve is fully closed, the compressor delivers no output. Some compressor manufacturers reduce the power required to drive this unloaded compressor by venting the output to the atmosphere (blow down), which reduces the zero output load from 50-70 percent FL to 20 percent FL. This procedure normally requires a separate heavy-duty oil pump to provide rotor lubrication.

Variable displacement rotary screw designs reduce the volume of air delivered to the output by venting a variable portion of the helical screw length to the inlet side of the compressor. One manufacturer uses several separate air controlled poppet valves along the screw while another uses a rotating turn valve that uncovers successive ports along the screw. These control features enable the compressor to operate at higher partial load efficiencies than do the inlet throttling valves. A continuous throttling range down to 40 percent capacity with a power requirement of 64 percent full load is achievable with a zero output power requirement of about 20 percent FL. Figure 4-9 compares the operating efficiencies of these three types of rotary screw compressors at various output levels.

A multiple compressor installation programmed so that some of the compressors can be operated at full capacity most of the time has been discussed elsewhere. The compressor manufacturers offer several different types of state-of-the-art control systems for such installations. Such multiple compressor/ control combinations should be considered and analyzed as alternates to one larger compressor operating partially loaded much of the time.

Once the power requirement and load profile of a factory area known, various compressor/control combinations can be analyzed for suitability to the requirement. Procedures such as those in Section 5.2.1 can be used for comparing the various options.

4.1.4 Aftercooler Efficiency

Similar cost analysis can be applied to aftercooler selection. Here there are several criteria:

Optimum size
Cooling medium
Maintainability
Control

As a general rule, oversizing aftercoolers is cost effective. Cooler discharge air has lower moisture content, easing the task of the dryer, and is more dense and hence will permit greater mass flow at the same system pressure drop. Also, larger coolers are likely to have less pressure drop themselves. One caution is that many aftercoolers are equipped with attached separators and care must be exercised to assure that the unit selected will not have a separator that is too large and, hence, ineffective.

Aftercoolers are cooled by either air or liquid. The liquid almost always is water. Usually, the larger units are liquid cooled but either can be effective. The temperature and quantity of available coolant will, of course, be key factors in selecting the type of aftercooler. In areas where water is costly, the cost of the water must be weighed against the cost penalties of warm compressed air. Larger systems can justify cooling towers and these are seen frequently in the southwestern states. Air cooled systems are more conveniently adaptable to heat recovery installations.

Higher costs can be justified for coolers that can be disassembled for cleaning. A dirty aftercooler costs in two ways: warmer air, and excessive air pressure drop. Always install a maintainable unit. This is particularly important for systems using lubricated compressors or high hydrocarbon content inlet air.

For best economy, coolant flow should be controlled. Where fresh water is used and then discarded or where the demand for cooling water exceeds re-use needs, the cost of a temperature control can be recovered, usually within one year. The temperature used as the basis for control can be outlet air temperature, outlet water temperature or operating approach temperature, the best of all. The modulating valve always should be installed on the water discharge side of the cooler. Precautions should be taken to handle any solids that settle out of the water.

4.1.5 Heat Recovery

Adiabatic compression of air to 100 psi results in outlet air temperatures of 350-500°F. When this air is cooled to ambient temperatures, 60-90 percent of the energy of compression is removed, and this can be used for other purposes. This heat is a low grade source that is available year round whenever the plant is in operation. Typical uses of the air include supplemental space heating, boiler makeup water preheating, or process heating.

If a regular all season use of this low grade heat source can be found, then heat recovery system efficiency considerations are of greater importance than the compressor efficiencies discussed in Section 4.1. Since nearly 80 percent of the input energy is available as heat, and if this heat can be used, then the 20 percent used to compress air is nearly a by-product. In multiple compressor systems, the compressors could be located near each of the areas requiring space heating to minimize the ducting requirements, air pressure control stability considerations permitting.

Several manufacturers market packaged air compressor/ heat recovery systems for air-cooled systems of greater than 50 HP and water-cooled systems greater than 125 HP. When designed as a complete system, the capital investment is low and return high. The cost usually is recovered in 1 to 3 years.

The rotary, helical screw, oil flooded compressor is most amenable to heat recovery for supplemental space heating in sizes greater than 50 HP. Heat exchangers and fans normally are incorporated in standard air compressor designs, to cool the oil and compressed air. The only modification required to incorporate heat recovery is suitable ducting to permit adjustable channeling of air from both inside and outside the building and a higher pressure distribution fan to handle the duct loading. When space heating is required, air is drawn from both inside and outside the building, heated and delivered to the heated spaces.

In summer, all heated air is exhausted unless there is a use for it elsewhere in plant operations. The energy available at full load typically is 56,000 BTU/hr per 100 cfm of compressor delivery, with achievable recovery efficiencies of 85-90 percent. This comes to approximately 50,000 Btu/hr per 100 cfm of recoverable energy at temperatures between 120 and 200°F. This type of heat recovery can only be able to supplement a building heat system for two reasons: The air compressor operates cyclically, varying from no load to full load during the day and is commonly off during the night when there may be heating requirements. Also, although the air compressor may consume 10 to 15 percent of the typical plant power, it may not be sufficient to supply the complete heating requirements.

Water cooled compressors are less amenable to heat recovery, due to the additional stage of heat exchange and the lower temperatures involved, reducing the overall recovery efficiency to 55-60 percent. It is only economically attractive for large compressors (greater than 125 HP) or regions with unusually high costs.

In estimating the value of a heat recovery system, it is important to note that 1 Btu of energy recovered can typically save 1.6 Btu of fuel required to produce the same Btu. This 1.6 × adjustment accounts for the combined effects of heating system efficiency and the heater part load correction factor. The part load correction factor treats the losses due to heating up and cooling down when cycling. This factor is especially applicable to the unit area heaters used so abundantly for factory heating.

It is important that the designer of a heat recovery system make himself knowledgeable in both air compressor operation and design and heating, ventilating and air conditioning (HVAC) design. Both disciplines have an impact on the design of a safe, reliable, and efficient heat recovery system. Improvised systems frequently are poorly engineered and ineffective and, at worst, could lead to air compressor damage due to overheating.

In performing cost analyses of compressor systems, heat

recovery should be included as an important option and this may alter the entire cost analysis. A realistic estimate should be made of the annual savings from heat recovery, making allowance for seasonal variations in the value of the recovered heat and the value of the fuel thus saved.

4.1.6 Filter Energy Efficiency

Two features are important to energy efficiency of compressed air filters: effectiveness and pressure drop. It is important to not over specify the air filtration requirements of a system since the air filter may provide the largest single point pressure drop in the system, sometimes 2-7 psi. Recall that at 100 psi each 1 psi lost in the distribution system results in 0.7 percent increased energy consumption at the compressor. Many plant air systems do not require oil free or highly filtered air. Those factories that have localized requirements for such air can use localized air filtration systems. When complete system air filtration is required, it is important to specify filter performance required, by maximum permitted particulate size or filter type: coarse, (40 μ), fine (5 μ), coalescing (0.1μ and 10 9/cm of hydrocarbons), or ultra high purity air (0.1μ and 1 ppm hc). Current coalescing filters are effective in removing oil and particulates, but most have a pressure drop of 3 to 5 psi when new and clean. The pressure drop increases with use, and maintenance should be scheduled as advised in Section 5.2.3.1.

It is desirable to oversize the air filter to reduce the pressure drop. A proper analysis, as per Section 5.2.1, will compare the life cycle energy cost of the pressure drop with the increased initial cost of the larger size filter. The most economic filter normally is larger than one selected by manufacturers' rated capacity. In making these comparisons, insist upon reliable pressure drop/ flow rate data.

Filter differential pressure gauges are a good investment for accurate determination of filter condition. The same can be said of filter differential pressure signals. An annoying sonic signal will command maintenance attention to a dirty filter.

The air intake filter is especially important because of the effect of even a slight drop in inlet air pressure on compressor efficiency. Each 1 percent decrease in suction pressure costs 1 percent in compressor mass flow output and efficiency. The inlet air filter should offer as small a pressure drop as possible and should be maintained diligently.

The local filter/lubricators also can be a source of considerable lost energy. If the pressure drop here is great, then tool efficiency is lost as shown in Figure 3-8. These filter/lubricators should be generously sized and well maintained.

4.1.7 Dryers

For reasons explained in Section 3.9, the operation of compressed air systems is improved if the air is at least partially dried. However, drying requires added costs and in most cases some increase in energy. Therefore, drying should be prescribed only to the extent necessary to the proper functioning of the compressed air system.

In determining how much drying is advisable, the first factor to consider is the lowest temperature that the compressed air distribution system will experience. Air having a dew point higher than the lowest system temperature will provide condensate with resulting rusting, scaling, distribution system deterioration, and pressure drop increase. The loss of air due to more frequently necessary line purging and blowdown will be an added cost.

If a compressor is equipped with a good aftercooler and separator, and the distribution system does not have lines outside the factory building, then the deliquescent type dryer may be adequate. This is the least expensive dryer to purchase and uses no energy, except for that associated with its pressure drop. It is simply a pressure vessel containing water absorbing chemicals which melt away in time. Using salt tablets normally can depress dew points by about 12 to 20 degrees and the use of potassium carbonate can depress the dew point by more than 30 degrees

Fahrenheit. The only significant operating costs are the materials and labor required for the chemicals. The deliquescent dryer does not function well if air is delivered to it at a temperature higher than 100° Fahrenheit. Therefore, if this type of dryer is used, the aftercooler should be adequately sized.

Filters should be installed downstream of this dryer in order to prevent any carry-over of salt into the system. Also, provision must be made for designing the systems to drain the chemical/ water mix and dispose of it. It is corrosive.

There are a number of situations where the performance of deliquescent type dryers is not adequate to system needs. Instrumentation, finishing, some processing, and other usages frequently dictate the availability of air at a lower dew point than can be delivered by a deliquescent dryer. Therefore, other dryers will be used in place of, or in series with, the deliquescent ones.

One of the more commonly used types is the refrigeration dryer and these can be specified to provide dew points as low as 35° Fahrenheit. The refrigeration type dryer has a pressure drop of about 3 to 5 psi at rated flow and, of course, requires some energy for operating the refrigeration plant. Reheating is recommended, so dryers with built-in reheaters, or the addition externally of reheaters, are advisable to return energy to the air and reduce condensate on the outside of the riser.

Power requirements for the refrigeration dryer typically will be about 6 percent of that of the compressor in a 100 psig system, assuming that all of the air from the compressor is to be dried. If the compressor is to be operated intermittently, then some economy can be achieved by cycling the dryer refrigerant compressor. This requires fast feedback controls, timing controls, or the addition of some thermal inertia or "fly-wheel effect" into the system. The latter can be achieved by having the refrigerant chill a thermal mass, in which both the air coils and refrigerant coils are embedded. In either case the refrigeration system then can cycle, more or less matching the load factor of the compressor. Such systems cost a bit more but, in cases where the requirements for dry air are variable, the savings in energy and maintenance costs could justify the higher initial cost. In some installations the frequent cycling may actually increase energy consumption, by comparison with continuous operation. Each case should be specifically analyzed.

The very driest air is produced by adsorption type or adsorption desiccant type dryers. These are similar to the deliquescent type except that instead of containing a chemical which melts away, they contain a reusable desiccant, usually silica gel, activated alumina or a molecular sieve. The silica gel type can provide dew points reliably to -40°F and sometimes even lower. The maximum inlet temperature should be about 100°F. The activated alumina type can accommodate temperatures to 120°F and can deliver dew points even lower.

It is important that any air passed through a desiccant type dryer be free of oil which will cause deterioration of the desiccant material. Slugs of water also are harmful to the dryer and should be avoided.

Normally, desiccant dryers consist of two vessels, each containing desiccant material. While one of the vessels is being used in a drying mode, the other is purged, with or without the addition of heat. Sometimes the purging is accelerated by a vacuum pump. Of these three purging methods, the vacuum type is the most energy efficient and the unheated purge type the most energy wasteful, the latter using 14 to 18 percent of total air flow just for purging. The electrically heated type uses 2 to 8 percent of the air for purging, but too high heating temperatures can lead to bed deterioration. There are now available new dryer designs which are much more energy efficient. These should be examined for all new and some older existing systems.

It is important to make an economic evaluation of how much drying is necessary. It has been found that reducing the dew point to just below any temperatures to be expected in the distribution system is a degree of drying which is relatively easy to justify. Drying to lower dew points is costly and justification will require more careful analysis. Much, of course, depends upon how the air is to be used and what the cost penalties of less dry air will be as this air affects the end use processes or devices. It is known, for example, that the operation of instrumentation needs dry air. Maintenance is reduced on tools and other pneumatic devices if the moisture is eliminated. However, the amount by which maintenance is reduced should be evaluated and compared to the added system initial and operating costs of the dryers. This comparison is not difficult to make and it requires only that all of the annual dryer costs, including power, pressure drop, operations, maintenance, purge, depreciation, and any other miscellaneous costs be summed and compared to the annual maintenance cost savings that will issue from dryer air.

In this way, the amount of drying that can be justified is determined and the dew point that is needed for operating the factory is specified. Do not overdry. If a portion of the factory requires especially dry air, furnish a dryer only for that portion. Some factories have two complete distribution systems, one supplying a large demand for dry air and the other a comparable need for less dry air, both to many dispersed applications.

Once the dew point and air quantities are agreed, investigate alternative dryers, considering first costs, operating costs, energy requirements, and maintenance costs, all from reliable manufacturers and experienced users. Keep in mind part load operating considerations. Complete a life time cost analysis of each dryer option as described in Section 5.2.1. Such an analysis is more than worth the effort.

4.1.8 Hoses and Connectors

The principle difficulty with hoses and connectors is that often they are undersized and prone to leakage. The leaks can be avoided by selecting quality products, installing them properly, and maintaining them. All new installations should be checked immediately and periodically thereafter. Quick connect type fittings are particularly vulnerable to environmental dirt when disconnected.

Referring to Figure 3-8, it can be seen that pressure losses should not be permitted at this point in the system. Reduced air pressure, and thus a less efficient tool, results in the use of more air to do the same job with more labor minutes expended.

When the application is for portable hand tools, the larger hoses and larger connectors, having lower pressure drop, are heavier and may result in decreased worker productivity. One approach is to install a large hose leading to the point where the operator is standing, and then transition to a shorter length of smaller hose leading to his tool.

To reduce the loss in connectors, consider using larger ones and omitting the quick connect ones wherever frequent discon-necting is not required. There are many factory installations which are semi-permanent and in which a quick connect fitting does not recover its cost in time saved, energy factors notwith-standing. Considering the additional energy and labor costs of frequently experienced connector pressure drops of 5 to 10 psi, and they can become costly indeed when they are not needed. Some designs are, however, much more efficient than others, so the pressure drop at the required flow rates should be investigated before purchase. In many cases improved operator efficiency can justify such couplings, especially more efficient ones. Each drop should be engineered for lowest lifetime cost.

4.1.9 Valves

The type of valves used throughout the system can make a difference. For example globe valves designed for water service are not appropriate for compressed air systems due to high pressure drop. The compressed air system designer should de-termine that the valves specified are designed for compressed air service. The valve should have:

- Minimum restriction and pressure drop when the valve is open
- No significant leakage through the stem during its service life, and

• Firm air shutoff in the closed position.

The materials used in the valve must be compatible with compressed air and any contaminants—especially lubricating oil, water, and trace gases.

The pressure drop across the valve in the open position is the most critical factor in compressed air systems. The pressure drop is a function of valve type and port size of the valve. Valve port sizes sometimes are lower than the pipe size by which the valve size is designated. The actual port size is the true "size" of the valve and must be used to determine pressure drop. Globe and angle valves have the highest pressure drop, while butterfly valves have much lower pressure drops. The gate, ball and plug valves are best.

Table 4-2 indicates the resistance to air flow for each valve type in the open position in terms of the equivalent length of pipe of the same diameter. This value can be related to a specific pressure drop. For example, a globe valve with a 90° stem has an L/D ratio (from Table 4-2) of 340. Such a valve with a 4" port then will offer resistance equivalent to $D \times L/D$, 4×340 inches, which is 113 feet of 4" pipe. For a flow of 1000 scfm at 100 psig, 100 feet of 4" pipe has a pressure drop of 0.19 psi; therefore, the pressure drop of the valve will be $113/100 \times 0.19 = 0.215$ psi. This pressure drop across the valve has an associated energy cost, for every year of operation, which is dependent on unit energy cost, compressed air system efficiency and operating time. A later section will indicate how to determine the most cost effective valve to use in a specific application. Care should be exercised in using and operating some "full flow" valves because they typically act quickly and can produce line pressure surges.

4.1.10 Regulators and Lubricators

Regulators and lubricators, when properly applied, can conserve energy in an air system. However, each of these devices represent a considerable pressure drop, typically 2 to 5 psid, and

Table 4-2 Typical Equivalent Length in Pipe Diameters of Various Valve Types

Valve Type		Equivalent length in pipe diameters, L/D
Globe -	90° Stem w/guide	450
	90° Stem	340
	60° stem	175
	45° Stem	145
Angle -	w/guide	200
	w/o guide	145
Check -	Conventional	135
	Clearway	50
Butterfly - 8" and larger		40
Cock -	Straight through	18
	Three way straight through	44
	-branch	140
Ball	5-10	
Gate	7-13	

Varigas Research, Inc.

should be used only where specifically indicated, the size and type carefully determined.

Regulators can result in savings when applied to devices that require a specific volume of air but do not require the full system pressure. Air cylinders frequently can be operated at reduced pressures, especially on a nonworking return cycle, at significant savings in air consumption. Regulators are frequently used to control accurately the output torque of a rotary tool, i.e., a screwdriver. Here the regulator should be sized to meet the desired pressure and flow requirements.

Lubricators frequently are recommended for air tools and cylinders. Sufficient and consistent lubrication can result in:

* Reduced air consumption for governed tools; also increased speed at the same air consumption for non-governed tools.

* Increased service life of rotor and cylinder seals.

* Reduced device maintenance cost or down time.

As mentioned elsewhere in this book, the regular, unplanned use of system-wide regulation can be wasteful. Some systems have been designed to compress at higher pressures than needed at the points of use, with later regulation to maintain constant pressure. A properly designed system, with generous sizing of the distribution network, should make this unnecessary in most cases. It is recommended that the system be so designed that compressor output pressure is as low as possible, consistent with end use requirements.

4.2 System Design

Given the characteristics of compressed air system components discussed in previous sections of this guidebook, it remains for the system designer to make intelligent, cost conscious, decisions directed toward the design of new systems or modifications to existing systems.

A consensus among experts in the compressed air industry is that the increased cost of energy has changed the way in which compressed air systems should be designed and specified. Many of the documents which comprise the bibliography for this report echo this sentiment. The major and most costly mistakes made by compressed air system designers are to buy larger compressors than are required or to add additional compressors to increase pressure in poorly operating systems. These mistakes can be avoided if the sources of energy loss are known and the correct design sequence is followed. Current guidelines for designing compressed air systems consist of the following steps:

1. prepare system layout
2. write system specification
3. size the distribution system
4. select compressed air conditioning equipment
5. determine compressor(s) size(es) and type(s)
6. select compressor driver(s)
7. select controls for compressor(s) and driver(s)
8. integrate energy-saving features
9. institute maintenance program

These steps should be followed when a new system is being designed and when additions or modifications are being made to an existing system. In the case of an existing system, a thorough analysis of the installed distribution system may lead to modifications which can provide the increased air requirement without requiring the purchase of additional air compression. This will save the first cost of the air compressor, but the cost recovery due to savings in energy usually will be greater.

4.2.1 System Layout

From the energy efficiency standpoint, system layout is the process of geometrically locating the air compressor, air piping and work stations for maximum plant efficiency and lifetime economy. Production variables, maintenance, location of uses, future capacity needs, power costs and energy consumption must all be considered in this process. Work stations and/or compressor locations might be selected to minimize the length(s) of lines between the air compressor and the largest user(s) of compressed air. In systems with a large distribution network, it might be preferable to have the compressor centrally located to minimize the farthest work station in the plant. The system should be arranged to minimize the number of valves, bends, transitions and other fittings and flow obstructions.

Several examples of compressed air system layouts are shown in Figure 4-10. Four types of headers, the loop, unit loop, grid and unit grid are shown in this figure. The loop header is the preferred arrangement and the grid header is the next most desirable. When the existing compressed air headers are too small or when there is not enough room to add additional compressors at the original locations, the unit loop and unit grid systems find application. The unit loop and unit grid systems usually are less energy efficient. When systems of this type are at part-load, there is normally more compressor capacity on line than necessary. More sophisticated control systems can help, not only with these layouts, but with others as well.

When multiple pressures are required for different plant operations, consideration should be given to separate distribution systems and compressors. But before deciding to install systems to provide air at other pressures, the annual usage requirement should be determined. If the demand for low pressure air is intermittent, it may not be economic to provide a separate system. Unless high pressure (greater than 100 psig) air is the major air usage in the plant, it is not economic to raise the entire plant air system to the higher pressure. Therefore, if some applications require higher pressure air, a separate high pressure system almost always is warranted.

The system layout, then, should determine four things:

• the number of systems required
• the locations of the work stations
• the compressor location(s)
• the piping arrangement(s)

Some other considerations in selecting the compressor location should be noted. Adequate space must be provided around

Figure 4-10   Distribution System Piping Diagram

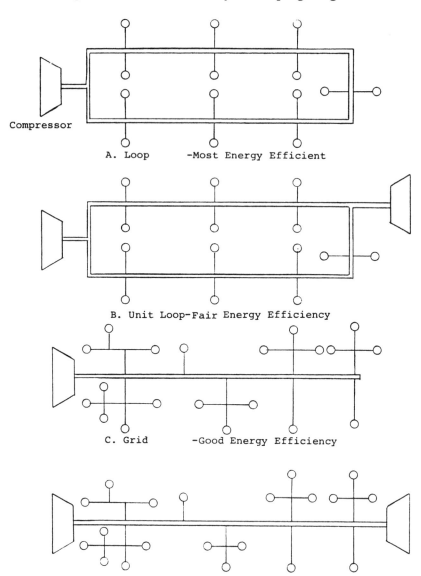

A. Loop   -Most Energy Efficient

B. Unit Loop-Fair Energy Efficiency

C. Grid   -Good Energy Efficiency

D. Unit Grid-Least Energy Efficient

the compressor for proper ventilation and for regular inspection and maintenance. Provision also should be made for coolants required. Experience has shown that, in general, compressors should be installed indoors in a heated compressor room with intake air ducted from the outside. Intake air should be taken from the coolest, cleanest and driest location available. The location also should be away from steam, chemical vapors and engine exhausts, and should be protected from the elements. An air filter on the intake is absolutely necessary to minimize dirt ingestion.

A variation on the compressor room installation has become possible with the introduction of integrated compressor/heat recovery packages. These permit placement of the compressor in the work area with waste heat used for space heating. However, the hot air may need disposal in summer months and cool air for compression must be ducted from outside.

For positive displacement compressors especially, cooler inlet air provides more efficient compressor operation. This is particularly true for the reciprocating and rotary screw compressors. Each 5° reduction in inlet air temperature provides approximately 1 percent more compresses air for the same input power. Looking at this information from the standpoint of energy cost, consider the following:

A 150 horsepower reciprocating compressor delivering 750 CFM with indoor intake (average temperature 70° Fahrenheit) has the intake moved outside in the shade where the average air temperature is 55°F. The compressor can deliver the same flow with 3 L percent less power. At 6¢ per kilowatt hour, this is a saving of $36 per shift per month (up to $1300/year). Furthermore, drawing air from a heated interior space in winter is wasting the energy that was used to heat the air.

One other factor to be noted about compressor location is that compressors are noisy. Many quiet factories would find them to be a terrible nuisance, so acoustic isolation, as in a compressor room, usually is indicated.

4.2.2. System Specification

The system specification is the basis for overall system design and component selection. It consists of a careful definition of the air flow, pressure, temperature, and quality at all critical points in the system. Collection of the operating data for the equipment to be supplied by the compressed air system is the first step in this process, which extends back through the distribution system, compressed air conditioning equipment and compressor discharge to the compressor inlet. The goal is to ensure that all factors which impact system operation and efficiency have been considered. This information will be used to size correctly the distribution system and compressed air conditioning equipment.

This process applies also to additions or expansions of existing compressed air systems. The designer must work his way back to the air requirement at the compressor installation and verify that the existing system has the capacity to handle the addition. This may involve re-analysis of the entire system.

System specification must consider four primary factors:

- the performance requirements of air operated equipment
- the operating environment of the compressor
- the availability of coolants which determines the method of cooling and the eventual temperature of the discharge air.
- future growth of the plant as it affects future air needs.

Table 4-3 is one format for collecting and presenting the data that comprise the air system performance requirements. These, in turn, determine the system specification. Since the compressor operating environment has a significant effect on performance, all flow data must be corrected from standard conditions (scfm) to the inlet operating conditions (icfm). Some compressor manufacturers prefer that the customer specify the desired output in scfm and the operating conditions: inlet pressure, temperature, humidity, cooling water temperature and pressure, and discharge pressure. The manufacturer will then select or design the com-

Table 4-3 System Performance Requirements

AIR USER REQUIREMENTS[1]

Process/ Branch #	Drop#/ Air User	Flow (icfm)[2]	Min. Press. (psig)	Mas. Temp. (°F)	Water Content	Oil Content

1. Enter values where specific requirements are known. Flow and pressure must always be specified. If values of air temperature, water content or oil content are unimportant, enter NS for not specified.

2. icfm specifies flow at compressor inlet conditions.

pressor to meet these conditions. The purchaser should ensure that the proper compressor size is specified. All equipment should be specified based on the highest flow and inlet temperature and lowest pressure to ensure that the desired performance is reached under all conditions.

The following terminology has been selected for describing the pipe components of the distribution system:

- The <u>discharge</u> <u>pipe</u> connects the compressor, aftercooler, aftercooler separator, receiver, filter and dryer. It usually includes a check valve and isolating valve. The <u>riser</u> (before distribution) connects the dryer (or last system component to the main distribution header.
- The <u>header</u> is the main distribution piping for the system.
- <u>Subheaders</u> supply air from the header to branch lines and drop lines.
- <u>Branch</u> <u>lines</u> supply groups of equipment from the subheaders.
- <u>Drop</u> <u>lines</u> supply individual equipment and work stations from the branch lines.

Consider a typical compressed air system layout, Figure 4-11. There are a number of work stations with varying functions and air requirements. Included are requirements for continuous low pressure air (25 psig) and intermittent higher pressure (150 psig) air. Following the recommendation of section 4.2.1, the decision has been made to provide separate 25 psig and 150 psig systems for those requirements.

A system specification for the 100 psig "plant air" system of Figure 4-11 is prepared in the following ten-step sequence:

1.   Identify each process or work station, the individual users at each station and the operating and performance data for each air consuming device. The required engineering data for a variety of air-powered equipment are listed in Table 4-4. Some of this information is dictated by the nature of plant

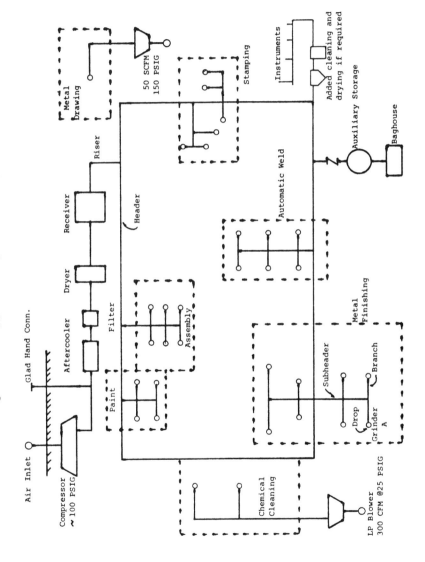

Figure 4-11 Typical Compressed Air System

operations. The remainder can be obtained from the equipment manufacturer's engineering data sheets.

2.   Assign a drop or feeder for each air user and define the performance requirements by specifying the flow, pressure, temperature and air quality at each drop. Also determine whether or not additional air conditioning is required at the point of use. For example, if paint spraying is done with plant air, a dedicated filter and dryer may be required.

Less than 2 psi of pressure drop (2 psid) should be allowed in the drop leg. One psid is a reasonable design goal. If hoses with standard fittings are used for portable tools, an additional 2 psid should be allowed. The use of quick-connecting couplings can add an additional 4 psid, while a lubricator will add 1 to 4 psid and a clean filter an additional 1/2 to 2 psid.

3.   Determine branch line requirements for the maximum or worst flow condition by summing the air requirements for all drops in that branch. Branch line pressure drop should be limited to 1 psid.

4.   Subheader capacity is then determined for the highest flow condition by summing the branch line requirements. Pressure loss in this part of the system should be limited to 1 psid.

5.   Determine header flow requirements by summing the products of air flow, duty cycle and work factor for all equipment in the system. Duty cycle is the percent time of use for each device while work factor is the percentage of full load flow actually used by the device. Continuous use equipment such as air cylinders and some assembly tools should derive the duty cycle from their frequency of operation expressed in cycles per minute.

Future growth in both plant size and production should

Table 4-4 Engineering Data for Air Powered Equipment

Engineering Data	Radial Vane Tools — Non-Governed: Ass'y Tools & Drills	Radial Vane Tools — Non-Governed: Impact Tools	Radial Vane Tools — Governed (e.g. Grinders)	Chippers & Scalers	Piston Air Motors	Cylinders	Valves	Blowguns	Paint Sprayers
Horsepower (max)	X	X							
Required Pressure (psig)	X	X	X	X	X	X	X	X	X
Air Consumption @ Max Rating & Required Pressure	CFM	CFM	CFM	CFM	CFM	CFC[1]	CFM[2]	CFM	CFM
CFM @ Free Speed & Required Pressure	X	X	X		X				
Torque Curve (rpm vs lb-ft)	X	X							
Stall Torque (lb-ft)	X	X			X				
Speed Curve (CFM to 120 psig)			X						
Inertia Rating (mr^2)		X				X			
Bore & Stroke Dimensions				X					
Blows per Minute				X					
Frequency of Operation	Duty Cycle (%)	Duty Cycle (%)	Duty Cycle (%)	Duty Cycle (%)	D.C. (%)	Cycles Min.	Cycles Min.	Duty Cycle (%)	Duty Cycle (%)
Work Factor[3]	X	X	X	X	X	X	X	X	X
Inlet Port Size	X	X	X	X	X	X	X	X	
Outlet Port Size						X	X		

1. Cubic feet per cycle.
2. Or required flow and velocity.
3. Percentage of actual flow to full load flow.

be factored into header design. Although it is not good practice to provide increased air compression until it is required, oversizing the distribution header and selected branch lines for two to five years of anticipated growth usually is a good investment.

Pressure drop in the header should be limited to 1 psid.

6. Riser flow requirement is the same as that of the header and pressure loss should be negligible if proper sizing recommendations are followed.

7. System leakage must be accounted for in sizing the air conditioning equipment and the compressor. Leakage ranges from 2 to 20 percent of system capacity, typically. The actual leakage is a function of maintenance; badly serviced systems can leak 30 to 40 percent of their capacity. Plant engineers should select a reasonable maximum permissible value such as 10 percent and institute a maintenance program whose goal is to maintain 5 percent or less.

8. The requirements for filtering and drying the compressed air must be specified based on the most demanding work station requirements where dedicated compressed air conditioning equipment is not installed. The pressure dew point at the dryer discharge should be 10 degrees lower than the lowest anticipated ambient temperature in the plant. This will probably occur during plant shutdown. The combined pressure drop of the filter and the dryer should be no more than 6 psid at a flow rate equal to header capacity, including system leakage. If a reheater powered by waste heat of the compressor is used downstream of the dryer, it is possible to reduce the capacity of the dryer and filter as well as the aftercooler and compressor.

Compressed air conditioning requirements must also

provide for removal of oil in liquid and aerosol form when lubricated compressors are planned.

Receivers also can be considered as part of the air conditioning system because some of the liquid contaminants will settle out there. Receivers normally are recommended with reciprocating compressors for removing pulsations from the air supply and for supplying a reserve volume to accommodate large transient requirements. Additional local receivers also are recommended in dedicated branch lines when supplying large volume transient users such as baghouses. When installed with reciprocating compressors, they frequently are located immediately downstream of the aftercooler. Care must be taken when sizing the filter and dryer and the distribution riser and header in this application. Presence of the receiver can allow flow transients which exceed compressor capacity and possibly the design capacity of the compressed air conditioning components. This would result in high pressure drop and poor air quality. A preferable location for the receiver may be after the compressed air conditioning equipment under these circumstances.

Specification of the receiver requires defining the operating pressure and volume. Only receivers which pass ASME code requirements and any state or local regulations should be used. The air compressor manufacturers should be consulted for guidance concerning matching receivers with their compressors. Guidance for sizing receivers to provide for transient high flow requirements is presented in Section 4.2.4 and for steady loads in Section 3.6.

9. The compressor and aftercooler requirements now can be specified. Selection of the proper compressor/aftercooler combination requires that system flow, pressure, temperature and oil and moisture content be specified.

Flow - The required flow at the aftercooler/compressor

discharge pipe is the header flow, plus leakage, plus the volume loss due to water removal in the dryer (purge air) and aftercooler, all expressed in icfm.

Pressure - Discharge pressure at the aftercooler or compressor discharge is determined by the maximum specified value of pressure loss plus operating pressure for the air powered equipment that is farthest from the compressor. For example, a 90 psig air grinder installed in the farthest drop from a compressor may require 92 psig in the branch line, 93 psig in the subheader and 94 psig in the header. With a 6 psi drop across the filter/dryer, the discharge pressure at the aftercooler must be 100 psig to power the grinder properly.

Temperature - The temperature of the air at the aftercooler/compressor discharge is influenced by several system parameters:

• the type of compressor being employed
• the presence or absence of an aftercooler
• the type of aftercooler
• the source of water for a water-cooled aftercooler
• the presence or absence of an air dryer
• the capacity of the air dryer

Some compressed air dryers require a maximum inlet temperature of 100°F. This requires an aftercooler with a 15°F approach operating with a maximum cooling medium temperature of 85°F.

A thorough consideration of these factors assures that selection of a discharge temperature is based soundly on the available utilities and the system requirements. For example, the unavailability of proper utilities for aftercoolers could influence the decision to favor oil-flooded screw compressors which are more often air cooled in medium and large sizes.

The temperature specification for this point in the system should include the aftercooler/compressor discharge temperature range (normally 90-100°F) and the approach temperature for the aftercooler. Remember the lower the approach and more expensive the aftercooler both in initial cost and operating expense.

Water and Oil - Water and oil are contaminants in the compressed air system. Their permissible content in the aftercooler/compressor discharge is a function of:

- the choice of compressor
- the capability of the downstream compressed air conditioning equipment to remove these contaminants
- the presence or absence of an aftercooler

The contaminant specification at this point should include the water content of the air, the particulate size in the air stream, and the oil carry-over in micrograms per unit volume.

The most important point to stress here is that too much conditioning of the air costs money. Appropriate specification is important in that it requires the designer to state the requirements in writing and then to design the system to provide only such conditioning of the air as is truly necessary.

10. Determine the compressor inlet air requirements. The performance of the air compressor is very much dependent upon the conditions at the compressor inlet. Therefore, the following information is essential to proper specification of the compressor:

- Inlet volume flow (icfm). This was determined in step 9 and requires knowledge of the range of both inlet air temperature and local barometric pressure
- availability of cooling water

- Compressor and air inlet locations
- Static pressure at the compressor inlet
- Range of relative humidity at the air inlet
- The presence of impurities in the inlet air. These include particulate, steam, chemical vapors, gases and engine exhaust.
- Availability of ventilation air in the compressor room
- Availability of electrical power or an alternate power source.

The information collected in these ten steps now permits complete specification of the compressor, after-cooler, filters and dryers and the sizing of the distribution system. This process may require some iteration, since the selection of several components is dependent on the performance of others. For example, final selection of the air dryer is dependent on the performance of the selected aftercooler and vice versa.

A partial specification for the compressed air system of Figure 4-11 is presented in Tables 4-5 and 4-6.

Table 4-5 System Performance Requirements
Air User Requirements[1]

Process/ Branch #	Drop#/ Air User	Flow (icfm)[2]	Min. Pres (psig)	Max. Temp. (°F)	Water Content	Oil Content[1]
Stamping/1	1/Cylinder	15	90	80	80°F pdp	NS
	2/Cylinder	15	90	80	80°F pdp	NS
	3/Cylinder	15	90	80	80°F pdp	NS
	4/Cylinder	25	90	80	80°F pdp	NS
Stamping/2	1/Cylinder	20	90	80	80°F pdp	NS
Auto Weld/1	1/Cylinder	12	90	80	80°F pdp	Lubricator Req'd
Metal Finish/1	1/Grinder	50	90	80	70°F pdp	Lubricator Req'd
	2/Grinder	25	90	80	70°F pdp	Lubricator Req'd

[1] Enter values where specified requirements are known. Flow and pressure must always be specified. If values of air temperature, water content or oil content are unimportant, enter NS for not specified.
[2] Flow at compressor inlet conditions.

Table 4-6   System Specification

System Component	Air Line			
	Max. Flow (icfm) In/Out	Min. Press. Max. (psig) In/Out	Allowable Water, Temp (°F) In/Out	Oil, hydrocarbons and Particulate
Subheader #1	70	93/92	100/100	No liquid water or oil
#2	110	93/92	100/100	No liquid water or oil
#3	45	93/92	100/100	No liquid water or oil
#4	85/85	93/92	100/100	No liquid water or oil
#5	120/120	93/92	100/100	No liquid water or oil
#6	50/50	93/92	100/100	Air to 10°F PDP, No liquid/aerosol oil particulate <0.01 micron.
#7	20/20	93/92	100/100	No liquid water or oil
Header		94/93	100/100	50°F PDP. No liquid/aerosol oil
Riser	500	94	100/100	No liquid/aerosol oil
Leakage	50	——	————	————————
Dryer	570/550	97/94	100/100	Saturated air, oil <10µg/ft3. Particulate <1 micron.
Filter	570/570	100/97	100/100	
Receiver	570/570	100	100/100	Not specified
Aftercooler/ Separator	580/570	106/100	280/100	Saturated
Compressor	580/580	/106	100/280	Inlet clean and dry

				Local Environment and Utilities		
Locations	Available Volts/Amps	Cooling Medium & Temp (°F)	Barometric Press Range (in. Hg.)	Relative Humidity Range (%)	Ambient Air Temp. Rangel °F	Inlet Air Content Particulate Vapors & Gases
Stamping				50-100		
Stamping				50-100		
Baghouse				50-100		
Auto Weld.				50-100		
Metal Finish.				50-100		
Paint				50-100		
Assembly				50-100		
All				50-100		
Comp. Rm.				50-100		
———				50-100		
Comp. Rm.	460/300	Air/100		50-100		
Comp. Rm.				50-100		
Comp. Rm.				50-100		
Comp. Rm.	460/300	Water 70		50-100		
Comp. Rm. Inlet Outdoors		Water 70	28.5 - 31.2	20-100	−10/100	None

Before proceeding to size the distribution system, the importance of providing for proper equipment function and performance should be stressed. The function of the compressed air system is to deliver the required power to the work stations. Money is wasted in both increased energy use and production losses if a compressor is selected which provides inadequate air supply at the equipment outlet. Costs also are increased if significant volumes of low pressure air are provided by reducing higher pressure air, although small or intermittent requirements for lower pressure air may best be provided by pressure reduction. It is important that the air system be optimally designed to provide proper equipment performance. Variations above and below design ratings contribute to equipment inefficiency and shorter equipment life.

The following observations are pertinent:

- Most air operated tools are designed to operate at 90 psig. Performance decreases 1.4 percent for every 1 psig decrease in pressure. Therefore, an air drill operated at 80 psig is 14 percent less effective than if operating at 90 psig.

- Performance can be increased by operating devices at higher than 90 psig, but the life of the device and accessory equipment is shortened. Such practice also may be a violation of safety regulations of the government (OSHA) or the company insurance policy.

- Operating equipment at pressures lower than 90 psig also contributes to excessive wear. This is particularly true of impact wrenches. Low pressure operation usually results in off-design performance of grinders and drills, in turn causing premature failure of grinding discs and drill bits.

- If it is not economically practical to provide 90 psig air at a particular location, an alternative is to select equipment that will provide the same performance at the lower pressure.

- If low pressure air is required for agitation or cleaning, centrifugal blowers or low pressure compressors can be used for less than half the power of using plant air. An example of this economy was observed in a small metal fabricating plant. A centrifugal blower recently was installed to provide 750 cfm of air at 25 psig to a bubbling system. The blower is driven by a 60 hp electric motor. By contrast, a two-stage reciprocating compressor would require 150 hp to provide that air flow through the plant air system and a pressure reducing valve. Since the discharge temperature of the air is not high, no conditioning is necessary and low cost plastic pipe can be used for transporting the air. All of these add up to a clear economic advantage for using blowers for low pressure air requirements.

4.2.3 The Distribution System

The basic goal of distribution system design is to deliver air to the end user with minimal system pressure loss, the consequences of which have been described. The distribution system therefore consists entirely of pipe, fittings, valves and hoses which connect the components.

Several basic guidelines should be applied to the design of all compressed air distribution systems:

- The discharge pipe from the compressor should be at least as large as the compressor discharge connection and it should run directly to the aftercooler.

- The riser, header, and selected branch piping should be oversized to permit 2 to 5 years of future growth in both plant size (new equipment) and production (higher duty cycle). This is cheap insurance against system obsolescence. Oversized pipe provides increased buffer storage and lower pressure drop and also requires fewer physical supports.

- The riser piping size and geometry are important for minimizing pressure drop and maximizing contaminant removal. A drip leg with a drain is necessary to prevent contaminants from flowing back to the compressor. The riser should be one size larger than the header and/or compressor discharge line. The horizontal leg should slope downward away from the header and should enter the header at the side or bottom. Unsuitable and recommended methods for joining the riser and header are shown in Figure 4-12.

- The header should be sloped 1/8" per foot away from the riser toward a low-point drain to facilitate condensate removal.

- The distribution and conditioning systems should be sized for a combined pressure drop of less than 10 percent of the compressor discharge pressure. A 100 psig compressor with a properly designed distribution and conditioning systems should therefore provide 90 psig air at the air powered equipment.

- A loop distribution system is recommended over the grid system because it provides two-way air flow to the points at greatest distance from the compressor. Piping should be sized to minimize pressure loss to the farthest outlet from the riser entry into the system.

- Fittings which offer the least flow resistance, such as long radius and street elbows, should always be used throughout the distribution system.

- Welded pipe and fittings are clearly superior and should be used wherever possible. Welded joints are less likely to leak and provide the least resistance to flow.

- Valves which offer the least flow resistance always should be used. Air leakage from valve stems should be minimized. This combination of requirements might dictate the use of

Figure 4-12 Examples of Riser Entry Configurations

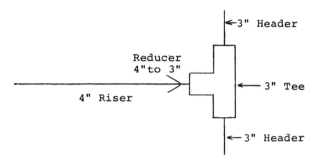

1. Poor Design- Significant pressure drop at riser.

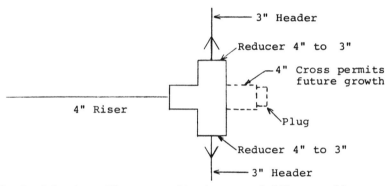

2. Good Design- Flow capacity increased 25% over #1.

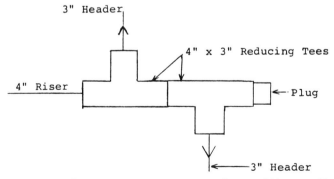

3. Best Design - Lowest pressure drop with growth potential.

ball and butterfly valves. However, the cost of ball valves justifies them only in difficult or critical applications. Gate valves have the highest potential for stem leakage but also the lowest pressure loss. They are most often specified, and should be maintained for low or no stem leakage.

- The subheaders, branch lines and drops should be as close as possible to the intended points of air us so as to minimize pressure loss in these smaller lines and hoses.

- Outlets from headers and branch lines should be taken from the top of the pipe to minimize water carry-over. Details of several examples are shown in Figure 4-13.

- Proper removal of moisture in drop lines requires a drain in the vertical leg. It is recommended that shutoff valves not be installed in the vertical leg but in a horizontal leg as shown in Figure 4-13.

- Use of hoses that are too long or too small in diameter, quick connect fittings and improperly sized filters and lubricators can significantly increase the pressure loss to the equipment. Guidance for hose selection is provided in Table 4-7.

Application of these guidelines and other techniques for assuring proper sizing and layout of the distribution system is one of the most neglected aspects of compressed air system design. It is, however, one of the most important because proper design can pay such high dividends in reduced energy costs over the life of the system. More importantly, it will provide for proper air supply at the point of use, resulting in good tool performance and optimal production at the work station. Increased energy consumption and lost production are the operating costs of a poorly designed distribution system. These easily can offset, within a few months, the lowered first cost of a poor system.

Parallel analyses of the distribution system can be performed for both performance and economy. However, the system

Figure 4-13 Recommended Piping Details
The Hansen Manufacturing Company

DETAIL — Branch Line to Drop

DETAIL — Subheader to Branch Line

DETAIL — Header to Subheader

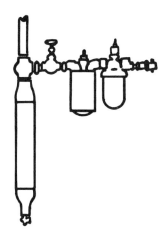

DETAIL — Drain Line   DETAIL — Drop Arrangement

Table 4-7 Pressure Drop for Air Supply Hoses

CFM Free Air Flowing	6'—1/8"	8'—5/32"	8'—1/4"	8'—5/16"	8'—3/8"	12½'—½"	25'—½"	50'—½"	25'—¾"	50'—¾"	8'—½" Plus 25'—½"	8'—½" Plus 50'—½"	12½'—½" Plus 25'—¾"	12½'—½" Plus 50'—¾"
2	3.5	1.2									1.3			
3	7.3	2.7									2.8			
4	12.5	4.4									4.6			
5		6.7									6.9			
6		9.3									9.7	1.2		
7		12.4	1.3								12.9	1.6		
8			1.6									2.1		
10			2.5									3.2		
12			3.5	1.3								4.5		
15			5.3	2.0				1.1				6.9		
20			9.0	3.4	1.4		1.0	1.9				11.8		
25			13.8	5.1	2.2		1.5	3.0					1.3	1.5
30				7.3	3.1	1.1	2.1	4.2					1.8	2.1
35				9.8	4.1	1.5	2.9	5.6					2.5	2.8
40				12.5	5.3	2.0	3.7	7.1		1.0			3.2	3.7
45					6.6	2.5	4.6	8.9		1.2			4.0	4.6
50					8.1	3.0	5.6	10.9		1.5			4.9	5.6
55					9.7	3.6	6.7	13.0		1.8			5.9	6.8
60					11.5	4.3	7.9		1.1	2.1			7.0	8.0
70						5.7	10.6		1.4	2.8			9.4	10.7
80						7.3	13.6		1.9	3.6			12.1	13.9
90						9.2			2.3	4.5				
100						11.2			2.8	5.5				
120									4.0	7.7				
140									5.4	10.3				
160									6.9	13.3				
180									8.7					
200									10.6					
220									12.7					

Based on 95 psiG air pressure at hose inlet end and includes normal couplings. Use of quick-connecting type couplings will increase pressure losses materially. The hose is assumed to be smooth.

Air is clean and dry. If an air line lubricator is used ahead of hose, the pressure loss will be considerably higher.

Pressure loss varies inversely as the absolute pressure (approximately).

Probable accuracy is believed to be plus-or-minus 10 percent.

Ingersoll-Rand

first should be analyzed for performance (i.e., minimum pressure loss). If economic justification is required for decisions to use oversize components, the procedures of Section 5.2.1 can be used.

Lines can be sized and pressure drops can be determined by

applying the empirical data shown in Tables 4-11 and 4-12. Note that two of the tables require prior knowledge of the compression ratio. This is simply the absolute pressure (gage pressure plus atmospheric pressure) divided by atmospheric pressure (14.7 psi). For example, if the line pressure is 95 psig, then the compression ratio is 7.46. Similar data are available from other sources. Care should be exercised in using such data because losses are not always expressed in the same units in the various publications. Sizing lines in this way is not difficult or especially time consuming, but it is an iterative process.

Computer programs are available which, of course, make such design much easier. These are especially easy to apply if a Computer Assisted Design (CAD) program is being used to lay out the distribution system. A calculator program has been developed also. Both programs are easier to use than the handbook technique since design data are stored for a variety of pipe and fittings. This makes data tables unnecessary for the most frequently used pipe and pipe fittings. The computer program will be menu driven to step the designer through the process and no knowledge of computers is necessary for utilizing this tool.

The preferred method for analyzing the requirements of the various piping components is to calculate requirements for the highest pressure equipment operating at the farthest point from the compressor. It is recommended that the pipe sizes which result from the analysis be used for all branches and drops. This technique reduces total system installed cost, provides for future growth and provides the flexibility for work station relocation within the system.

Several examples of the design procedure can be presented by continuing with the compressed air system example presented in Figures 4-11 and Tables 4-5 and 4-6. Analysis is presented for grinder A in the metal finishing department which is the farthest tool from the compressor connection to the loop. A potential piping arrangement for the drop line is shown in Figure 4-14. The branch line pressure of 92 psig is chosen to be consistent with the system specification of Section 4.2.2. Using the grinder flow of 50 icfm predicts a pressure at the tool inlet of 91 psig. The results of the computer pressure 1089 computation for these configurations are presented in Table 4-8 Addition of 25 feet of 3/4" hose from Table 4-7 to the grinder drop adds a 1.4 psi loss. This provides an acceptable 89.6 psig operating pressure at the grinder inlet. Addition of quick connect couplings and a filter/regulator/lubricator would reduce operating pressure to a marginal level.

Figure 4-14   Piping Configuration for Grinder Drop
Varigas Research, Inc.

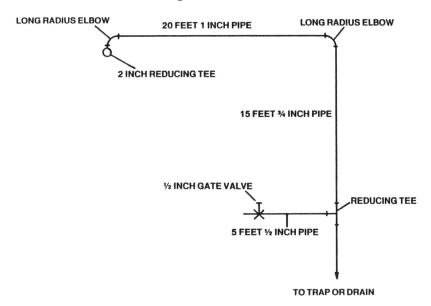

Analyses can also be performed of header configurations for this same compressed air system. Potential designs for the header are presented in Figure 4-15. Maximum flow, including growth,is 500 icfm at an entry pressure of 94 psig. The computer analysis predicts operating pressures at the metal finishing department subheader of 93.4 for configuration A and 91.4 for Configuration B. These results are presented in Tables 4-9 and 4-10.

Figure 4-15 Compressed Air System Header Design
Varigas Research, Inc.

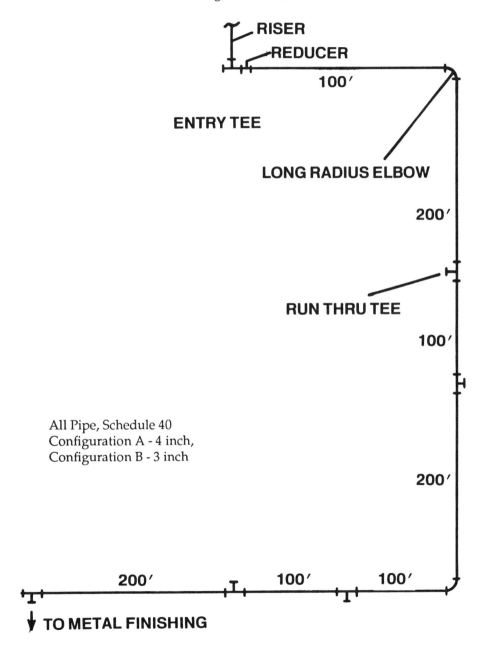

RISER

REDUCER

100′

ENTRY TEE

LONG RADIUS ELBOW

200′

RUN THRU TEE

100′

All Pipe, Schedule 40
Configuration A - 4 inch,
Configuration B - 3 inch

200′

200′   100′   100′

TO METAL FINISHING

Branch lines for transient high volume flow are designed separately. The subheader is sized to supply a receiver which in turn provides the capacity for transient peaks. The receiver is sized using the expression

$$V = \frac{T\, C\, P_o}{(P_1 - P_2)} \qquad \text{Equation (4-1)}$$

Where:

V = receiver volume in cubic feet
C = air flow requirement in scfm
po = atmospheric pressure in psia
$p1$ = initial receiver pressure in psig
$p2$ = final receiver pressure in psig
T = operating time of the air user in minutes

This calculation results in a conservative estimate of receiver size because it assumes no flow to the receiver while it is being discharged.

Table 4-8 Pressure Loss Calculation for Grinder Drop Station
Varigas Research, Inc.

No. 1
2" Reducing Tee: 1/2
50
0 psid
92 psig

No. 2
l" Elbow: Long Radius
50 scfm
01 psid
92 psig

No. 3
l" Pipe
20 ft
50 scfm
13 psid
91.9 psig

No. 4
l" Elbow: Std. 90
50 scfm
02 psid
91 8 psig

No. 5
3/4" Pipe
15 ft
50 scfm
.32 psid
91 5 psig

No. 6
3/4" Reducing Tee: 1/4
50 scfm
.04 psid
91 5 psig

No. 7
1/2" Pipe
5 ft
50 scfm
.45 psid
91 psig

No. 8
1/2" Gate Valve
50 scfm
.06 psid
91 psig

Table 4-9—Pressure Loss Calculation for 4 inch System Header—Configuration A

No. 1
5' Tee: Std. Side In/Out
500 scfm
0 psid
94 psig

No. 2
5' Reduction: 3/4
500 scfm
0 psid
94 psig

No. 3
4" Pipe
200 ft
500 scfm
11 psid
93.9 psig

No. 4
4" Elbow Long Radius
500 scfm
0 psid
93.9 psig

No. 5
4" Pipe
200 ft
500 scfm
11 psid
93.8 psig

No. 9
4" Pipe
200 ft
11 psid
93.6 psig

No. 10
4" Tee- Std. side In/Out
500 scfm
.01 psid
93.6 psig

No. 11
4" Pipe
100 ft
500 scfm
.06 psid
93.5 psig

No. 12
4" Tee: Std. Run
500 scfm
0 psid
93.5 psig

No. 13
4" Pipe
100 ft
500 scfm
.06 psid
93.5 psig

Table 4-9 Continued

No. 11
4" Pipe
100 ft
500 scfm
0.06 psid
93.5 psig

No. 12
4" Tee: Std. Run
500 scfm
0 psid
93.5 psig

No. 13
4" Pipe
100 ft
500 scfm
0.06 psid
93.5 psig

No. 14
4" Tee: Std. Run
500 scfm
0 psid
93.5 psig

No. 15
4" Pipe
200 ft
500 scfm
0.11 psid
93.4 psig

Table 4-10—Pressure Loss Calculation for 3-inch System Header—Configuration B

No. 1
4" Tee: Std. Side In/Out
500 scfm
0 psid
94 psig

No. 2
4" Reduction: 3/4
500 scfm
0.01 psid
94 psig

No. 3
3" Pipe
100 ft
500 scfm
0.23 psid
93.8 psig

No. 4
3" Elbow, Long Radius
500 scfm
0.01 psid
93. 7 psig

No. 5
3" Pipe
200 ft
500 scfm
0.46 psid
93.3 psig

No. 6
3" Tee: Std. Run
500 scfm
0.01 psid
93.3 psig

No. 7
3" Pipe
100 ft
400 scfm
0.23 psid
93.1 psig

No. 8
3" Tee: Std. Run
500 scfm
0.01 psid
93.0 psig

No. 9
3" Pipe
200 ft
500 scfm
0.46 psid
92.6 psig

No. 10
3" Elbow, Long Radius
500 scfm
0.01 psid
92.6 psig

Table 4-10 Continued

No. 11
3" Pipe
100 ft
500 scfm
0.23 psid
92.3 psig

No. 12
3" Tee: Std. Run
500 scfm
0.01 psid
92.3 psig

No. 13
3" Pipe
100 ft
500 scfm
0.23 psid
92.1 psig

No. 14
3" Tee: Std. Run
500 scfm
0.01 psid
92.1 psig

No. 15
3" Pipe
200 ft
500 scfm
0.46 psid
91.6 psig

Table 4-11

Friction of Air in Pipes

Divide the Number Corresponding to the Diameter and Volume by the Ratio of Compression. The Result is the Loss in Pounds per Square Inch in 1,000 Feet of Pipe.

Cu. Ft Free Air per Min.	Nominal Diameter in Inches								
	½	¾	1	1¼	1½	1¾	2	2½	3
5	8.9	2.0	.5	...	...	...	...	...	...
10	35.4	8.0	2.2	.5	...	...	...	...	...
15	79.7	17.9	4.9	1.1	...	...	...	...	...
20	142.	31.8	8.7	2.0	.9	...	...	...	...
25	221.	49.7	13.6	3.2	1.4	.7	...	...	...
30	318.	71.	19.6	4.5	2.0	1.1	...	...	...
35	434.	97.5	26.6	6.2	2.7	1.4	...	...	...
40	567.	127.	34.8	8.1	3.6	1.9	...	...	...
45	716.	161.	44.0	10.2	4.5	2.4	1.2	...	...
50	885.	199.	54.4	12.6	5.6	2.9	1.5	...	...
60	.....	286.	78.3	18.2	8.0	4.2	2.2	...	...
70	.....	390.	106.6	24.7	10.9	5.7	2.9	1.1	...
80	.....	510.	139.2	32.3	14.3	7.5	3.8	1.5	...
90	.....	645.	176.2	40.9	18.1	9.5	4.8	1.9	...
100	.....	796.	217.4	50.5	22.3	11.7	6.0	2.3	...
110	.....	963.	263.	61.1	27.0	14.1	7.2	2.8	...
120	.....	.....	318.	72.7	32.2	16.8	8.6	3.3	...
130	.....	.....	369.	85.3	37.8	19.7	10.1	3.9	1.2
140	.....	.....	426.	98.9	43.8	22.9	11.7	4.6	1.4
150	.....	.....	490.	113.6	50.3	26.3	13.4	5.2	1.6
160	.....	.....	570.	129.3	57.2	29.9	15.3	5.9	1.9
170	.....	.....	628.	145.8	64.6	33.7	17.6	6.7	2.1
180	.....	.....	705.	163.3	72.6	37.9	19.4	7.5	2.4
190	.....	.....	785.	177.	80.7	42.2	21.5	8.4	2.6
200	.....	.....	870.	202.	89.4	46.7	23.9	9.3	2.9
220	.....	.....	.....	244.	108.2	56.5	28.9	11.3	3.5
240	.....	.....	.....	291.	128.7	67.3	34.4	13.4	4.2
260	.....	.....	.....	341.	151.	79.0	40.5	15.7	4.9
280	.....	.....	.....	395.	175.	91.6	46.8	18.2	5.7
300	.....	.....	.....	454.	201.	105.1	53.7	20.9	6.6

Cu. Ft Free Air per Min.	Nominal Diameter in Inches								
	2	2½	3	3½	4	4½	5	6	8
320	61.1	23.8	7.5	3.5	...	...	...	...	...
340	69.0	26.8	8.4	3.9	2.0	...	...	...	...
360	77.3	30.1	9.5	4.4	2.2	...	...	...	...
380	86.1	33.5	10.5	4.9	2.5	...	...	...	...
400	94.7	37.1	11.7	5.4	2.7	...	...	...	...
420	105.2	40.9	12.9	6.0	3.1	...	...	...	...
440	115.5	44.9	14.1	6.6	3.4	...	...	...	...
460	125.6	48.8	15.4	7.1	3.7	2.0	...	...	...
480	137.6	53.4	16.8	7.8	4.0	2.2	...	...	...
500	150.0	58.0	18.3	8.5	4.3	2.4	...	...	...
525	165.0	64.2	20.2	9.4	4.8	2.6	...	...	...
550	181.5	70.2	22.1	10.2	5.2	2.9	...	...	...
575	197.	76.7	24.2	11.2	5.7	3.1	...	...	...
600	215.	83.5	26.3	12.2	6.2	3.4	...	...	...
625	233.	92.7	28.5	13.2	6.8	3.7	...	...	...
650	253.	98.0	30.9	14.3	7.3	4.0	2.2	...	...
675	272.	105.7	33.3	15.4	7.9	4.3	2.4	...	...
700	294.	113.7	35.8	16.6	8.5	4.6	2.6	...	...
750	337.	130.5	41.1	19.0	9.7	5.3	2.9	...	...
800	382.	148.4	46.7	21.7	11.1	6.1	3.3	...	...
850	433.	168.	52.8	24.4	12.5	6.8	3.8	...	...
900	468.	189.	59.1	27.4	14.0	7.7	4.2	...	...
950	541.	209.4	65.9	30.5	15.7	8.6	4.7	...	...
1000	600.	232.0	73.0	33.8	17.3	9.5	5.2	1.9	...
1050	658.	256.	80.5	37.3	19.1	10.4	5.8	2.1	...
1100	723.	280.6	88.4	40.9	21.0	11.5	6.3	2.4	...
1150	790.	306.8	96.6	44.7	22.9	12.5	6.9	2.6	...
1200	850.	344.0	105.2	48.8	25.0	13.7	7.5	2.8	...
1300	.....	392.0	123.4	57.2	29.3	16.0	8.8	3.3	...

Table 4-11 (Conclusion)

Friction of Air in Pipes

Divide the Number Corresponding to the Diameter and Volume by the Ratio of Compression. The Result is the Loss in Pounds per Square Inch in 1,000 Feet of Pipe.

Cu. Ft. Free Air per Min.	Nominal Diameter in Inches							
	3½	4	4½	5	6	8	10	12
1400	66.3	33.9	18.6	10.2	3.8	...	...	...
1500	76.1	39.0	21.3	11.8	4.4	...	...	...
1600	86.6	44.3	24.2	13.4	5.1	...	...	...
1700	97.8	50.1	27.4	15.1	5.7	...	...	...
1800	110.0	56.1	30.7	16.9	6.4	...	...	...
1900	122.	62.7	34.2	18.9	7.1	1.6	...	...
2000	135.	69.3	37.9	20.9	7.8	1.8	...	...
2100	149.	76.4	40.8	23.0	8.7	2.0	...	...
2200	166.	83.6	45.8	25.3	9.5	2.2	...	...
2300	179.	91.0	50.1	27.6	10.4	2.4	...	...
2400	195.	99.8	54.6	30.1	11.3	2.6	...	...
2500	212.	108.3	59.2	32.6	12.3	2.9	...	...
2600	229.	117.2	64.0	35.3	13.3	3.1	...	...
2700	247.	126.	69.1	38.1	14.3	3.3	...	...
2800	265.	136.	74.3	41.0	15.4	3.6·	...	...
2900	285.	146.	79.8	43.9	16.5	3.9	...	...
3000	305.	156.	85.2	47.0	17.7	4.1	...	...
3200	347.	177.	97.1	53.5	20.1	4.7	...	...
3400	391.	200.	109.5	60.4	22.7	5.3	...	...
3600	438.	224.	122.8	67.7	25.4	5.6	1.8	...
3800	488.	250.	137.	75.5	28.4	6.6	2.0	...
4000	542.	277.	151.	83.6	31.4	7.3	2.2	...
4200	....	305.	168.	92.1	34.6	8.1	2.4	...
4400	....	335.	183.	101.2	38.1	8.9	2.7	...
4600	....	366.	200.	110.5	41.5	9.7	2.9	...
4800	....	399.	218.	120.4	45.2	10.5	3.2	...
5000	....	433.	236.	131.	49.1	11.5	3.4	...
5250	....	477.	260.	144.	54.1	12.6	3.8	...
5500	....	524.	286.	158.	59.4	13.9	4.2	1.6
5750	....	....	313.	173.	64.9	15.2	4.6	1.8
6000	....	....	341.	188.	70.7	16.5	5.0	1.9
6500	....	....	402.	222.	82.9	19.8	5.9	2.3
7000	....	....	464.	256.	96.2	22.5	6.8	2.6
7500	....	....	532.	294.	110.5	25.8	7.8	3.0
8000	....	....	....	335.	125.7	29.4	8.8	3.6
9000	....	....	....	423.	159.	37.2	11.2	4.4
10000	....	....	....	523.	196.	45.9	13.8	5.4
11000	....	....	....	....	237.	55.5	16.7	6.5
12000	....	....	....	....	282.	66.1	19.8	7.7
13000	....	....	....	....	332.	77.5	23.3	9.0
14000	....	....	....	....	387.	89.9	27.0	10.5
15000	....	....	....	....	442.	103.2	31.0	12.0
16000	....	....	....	....	503.	117.7	35.3	13.7
18000	....	....	....	....	636.	148.7	44.6	17.4
20000	....	....	....	....	....	184.	55.0	21.4
22000	....	....	....	....	....	222.	66.9	26.0
24000	....	....	....	....	....	264.	79.3	30.1
26000	....	....	....	....	....	310.	93.3	36.3
28000	....	....	....	....	....	360.	108.0	42.1
30000	....	....	....	....	....	413.	123.9	48.2

Tables on pages 105 to 110 based on Formula of E. G. Harris Univ. of Mo., bulletin No. 4, Vol. 1, 1912.

$$D = \frac{CL}{r} \times \frac{Q^2}{d^5} \text{ in which } C = \frac{.1025}{d^{0.81}} \text{ for ordinary pipe.}$$

$$D = \frac{.1025 L Q^2}{r d^{5.81}} \text{ in which}$$

D = pressure drop in lbs. per sq. in.
L = length of pipe in feet.
Q = cu. ft. of free air per sec.
r = ratio or compression at pipe entrance.
d = internal diameter of pipe in inches.
C = experimental coefficient.

From "Compressed Air Data" Pub. by Compressed Air Magazine

Piping on the down-stream side of the auxiliary receivers should be sized to handle the peak flow, C, expressed in scfm, while the subheader piping upstream from the receiver should be sized to resupply the receiver during off cycles. Auxiliary compressed air conditioning equipment, if required, normally is installed between the receiver and the main header with a check valve between them to prevent back flow.

Design of the distribution system also should incorporate features to minimize system leakage. For example,

Table 4-12

Representative Equivalent Length in Pipe Diameters (L/D) **of Various Valves and Fittings**

Globe valves, fully open	450
Angle valves, fully open	200
Gate valves, fully open	13
¾ open	35
½ open	160
¼ open	900
Swing check valves, fully open	135
In line, ball check valves, fully open	150
Butterfly valves, 6 in and larger, fully open	20
90° standard elbow	30
45° standard elbow	16
90° long-radius elbow	20
90° street elbow	50
45° street elbow	26
Standard tee:	
Flow through run	20
Flow through branch	60

Compiled from data given in "Flow of Fluids," Crane Company Technical Paper 410, ASME, 1971.

• Valves should be used to isolate portions of the plant that are used seasonally or are shut down for weeks or months at a time. These may be electrically operated valves, wired in parallel with the electrical power to the machines in that portion of the plant.

• Electrically operated valves also can be used to isolate the air supply to machines when they are not in use. This is particularly applicable to machines which are prone to many leaks or have a history of bad maintenance.

Economic analyses can be performed on the various components in the distribution system to balance the increased cost of a more efficient system against the resulting power cost savings. The method described in Section 5.2.1 may be helpful.

4.2.4 Compressed Air Conditioning Equipment

Compressed air conditioning equipment in a compressed air supply system must provide both for contaminant removal and for preparation of the air for equipment use. The compressed air conditioning system which provides for all of these functions includes the following:

- a filter for the compressor inlet air

- an aftercooler to reduce air temperature and remove moisture with a separator for removing liquid water

- a receiver to reduce air pressure pulsation and moisture carry-over; also to decrease cycling in small systems

- a dryer for removing water vapor and liquid

- a filter or coalescer to remove water, oil, and particulates from the air stream

- traps or drains for draining water from the air stream with minimum air loss

- lubricators to provide equipment with proper lubrication

- silencers for reducing noise from the compressor inlet and air line blowoffs

- regulators to provide reduced,regulated air pressure

The energy efficiency of a compressed air system can be enhanced by the proper application of these components since they are not all necessary components to every system and since most of them remove energy from the air.

Maximizing the energy efficiency of the compressed air system starts in the design process by selecting the best item for each application. This must be followed by proper installation and proper maintenance. Factors which affect the selection of the optimum air conditioning equipment include:

Air Use—The ultimate application of the compressed air is the primary criterion. The quality of compressed air for manufacturing operations is not as critical as for the food, pharmaceutical and textile industries.

In manufacturing, air used for cleaning and soot blowing requires little conditioning, while air for powering tools and machinery and for painting require varying amounts of conditioning. Serious consideration should be given to these different requirements and,if feasible, separating the low grade uses prior to air conditioning equipment. This saves both equipment and energy costs. Some factories distribute both conditioned and unconditioned air.

Climate—Air line fouling and freezing are common problems in industrial compressed air systems. The most economical method for preventing these problems is drying compressed air to a pressure dew-point that is 10 below the lowest expected ambient temperature. Dry air also improves pneumatic equipment performance.

Plant Layout—The relative locations of work stations and the air compressor have a major impact on the type of air conditioning equipment selected. The system configuration can often dictate whether the conditioning equipment should be located near the compressor or should be distributed through the system. Maintenance should be considered when deciding on placement of this equipment.

Inlet Air—The quality of the air entering the inlet of the compressor determines the amount of filtration required and the location from which the air is taken. The design of the compressor inlet can have a significant impact on the energy efficiency of the compressed air system.

Water Availability and Cost—This factor determines whether or not air or water is used as the cooling medium for the aftercooler.

Water Disposal and Sewer Costs—Disposal of oily waste water from the compressed air auxiliary equipment is a growing problem in terms of pollution control and cost. The rapidly increasing sewer surcharges placed on industrial water users have doubled the cost of cooling water in many communities.

Equipment Arrangement—Some guidelines for the proper placement of compressed air auxiliary equipment have been given in other sections of this book. To summarize the preferred arrangement for this equipment: The inlet air filter is the first item seen by the incoming air. The aftercooler and separator should be located in the compressor discharge as close as reasonable to the compressor. The remainder of the system arrangement is dependent on the particular application. The preferred arrangement for systems with sudden demand cycles is: filter, dryer, then receiver to prevent surges in the system from exceeding the dryer and filter air flow capacity. However, the majority of systems with a steady air flow install the receiver upstream from the dryer to reduce pressure pulsation. Drains and traps should be installed on all system components, including the separator, filters, dryer and receiver, at low points in the overhead distribution piping and in all drop legs of the distribution piping.

EQUIPMENT SELECTION

The system specification provides guidance for selecting the individual conditioning components. Since most components are rated in scfm, it will probably be necessary to calculate flow in both icfm and scfm. Selection then should be a matter of weighing performance versus total cost which includes the purchase price, operating (including energy) costs and maintenance costs.

Inlet Air Filters—Since there is a considerable amount of dirt in even the cleanest atmospheric air, some filtration of compressor inlet air is always recommended. The primary function of these filters is to remove particulates as small as three microns with minimum pressure loss. There are three primary types of inlet air filters: dry, oil wetted, and oil bath. The performance of these filters is summarized in Table 4-13. The dry filter has the highest filtration efficiency and a low pressure drop. It also is compatible with all air compressors including non-lubricated designs. Maintenance consists simply of replacing the element. This is, therefore, the filter of choice except for very high, steady flow situations. Multistage filters or combinations of filters in series can be used for severe filtration requirements.

The pressure drop for the dry filter increases dramatically with the addition of some silencers. Use of a separate silencer is therefore recommended only when justified, for example, reciprocating compressor intakes that are indoors or near an office window or a local property line.

Aftercoolers—An aftercooler usually is recommended for all installations because it is an economical way of reducing the moisture content of the air. However, some compressors can function without a separate aftercooler, particularly oil flooded screw compressors which have a discharge temperature of 100°F above the ambient. This will result in a great deal of water in the distribution system.

The decision of whether to use water or air-cooled aftercoolers is based on four factors. In order of importance they are:

• feasibility of employing heat recovery system
• the availability, cost and temperature of cooling water
• the required compressed air flow and temperature
• the ambient air temperature and ventilation conditions

Regardless of the cooling method chosen, a compressed air by-pass line and valve should be installed to permit servicing the aftercooler.

Table 4-13 Inlet Air Filter Performance
Varigas Research, Inc.

Filter Type	Filter Action Efficiency (%)	Particle Size (Microns)	Max Clean Δp (in H$_2$O)	Comments
Dry	100 99 98	10 5 3	3-8	1.
Viscous- Impingement (Oil Wetted)	95 85	20 10	0.25-2.0	2.,3.
Oil Bath	98 90	10 3	2.0 lowΔ p 6-10	2.,3.,4.
Dry with Silencer	100 99	10 5	5.0 [5] 7.0 [6]	1.

1. Recommended for non-lubricated compressors and for rotary vane compressors in a high dust environment.
2. Not recommended for dusty areas or for nonlubricated compressors.
3. Performance requires that oil is suitable for both warm and cold weather operation.
4. Recommended for rotary vane compressors in normal service.
5. Full flow capacity up to 1600 cfm.
6. Full flow capacity from 1600 cfm to 6500 cfm,

Water cooled units provide better heat transfer and require no electrical power. Therefore, they are most often recommended if water cost is not excessive. Air cooled aftercoolers can be installed outdoors if a clean area is available near the compressor. Outdoor installed air cooled units must be protected from cold weather freezing of condensate by a thermostatically controlled heater. Air-cooled aftercoolers are more easily applied to heat recovery. Since 65 to 80 percent of the input power can be made available as heat, the recovery efficiency of a given candidate system and aftercooler can be the primary design factor. See sections 4.1.4 and 4.1.5.

The following recommendations apply to most water cooled installations:

- Some aftercoolers have a significantly high pressure drop (3 to 5 psi) at rated flow; therefore it is good economy to oversize the aftercooler.

- Reduce water and sewage costs by installing modulation control on the cooling water supply. This is especially recommended for oversized coolers. The control signal should be supplied by the discharge air temperature or, alternatively, the discharge water temperature.

- Water flow throttling should be done on the downstream or water discharge side of the aftercooler. This technique keeps the aftercooler full of water and under pressure.

- The aftercooler functions best when installed indoors in a horizontal position. Space should be provided for pulling and cleaning the tubes.

Separators—A separator usually is supplied with an aftercooler, and for this reason it is important to specify the design air flow for the system, particularly when using an oversized aftercooler. Separators are designed to match specific flow conditions and should not be oversized. Since the pressure drop across a properly sized separator is approximately 3 psi, a poorly specified separator which is not providing the proper function is wasting 2 percent of the input to the system.

Separators should always be equipped with drains or traps for removing condensate.

Receivers—Receivers normally are sized for about 10 percent of the compressor capacity in cfm. However, it is best to discuss the selection of receiver with the compressor manufacturer. Receivers are not required in all installations; however, they are normally recommended for use with reciprocating compressors to remove the pressure pulsations from the air stream. They also can provide energy savings by decreasing the cycle frequency in systems using start/stop control. This is especially advantageous in small systems with little inherent storage capacity. This technique saves motor starting energy.

Air entry-to the receivers should be below the center, and the discharge should be near the top to minimize carry-over of entrained liquids. The receiver always should be an ASME design pressure vessel with a safety valve installed. The large plug fittings on receivers are prone to leaks and should be thoroughly checked at installation.

Dryers—The system specification has determined the plant-wide ambient air temperatures. These should be used as a guide in specifying the performance of the compressed air dryer. To prevent moisture from condensing in the distribution system, the pressure dew point of the air discharging from the dryer should be 10 degrees lower than the lowest expected ambient temperature in the factory.

Installation arrangements and performance specifications for compressed air dryers are presented in Figure 4-16 and Table 4-14 respectively. Using this information, the dryer installation should be carefully designed to provide just the amount of drying necessary. Oversizing dryers is not recommended. If dry air is required at particular work stations, then point-of-use drying equipment can be installed. If the required dew point for the point-of-use applications is known, the required discharge pressure dew point of the dryer can be determined using Figure 4-17.

Economic analyses to support dryer selection can be performed using the technique described in Section 5.2.1.

It is important to consider total energy cost when calculating the cost comparison of dryers. Energy costs must include:

- Cost of electricity or steam
- Cost of replacing solid hygroscopic desiccants
- Pressure drop across the drying system, including filters
- Cost of purge air

Table 4-14 Compressed Air Dryer Performance

Dryer Type	Inlet Air Capacity @ 100°F				Outlet Air @ 100°F Ambient				Power		Prefilter(4) Afterfilter	Installation
	Flow (scfm)	Press (psig)	Max. Temp. (°F)	Moisture (%rh)	Press (psig)	Moisture (°Fpdp)	Cooling	Req'd				
Deliques-cent	5-30,000	100	100	Saturated	95	80	None	None. Requires drying medium replenishment			Recommended Required (1).	Indoor & Outdoor
Refriger-ated	5-25,000 (2)	100	130	Saturated	95	35 to 50	Air @ 100°F or Water @ 85°F	Electrical (3)			Recommended Not Required	Indoor
Desiccant Regenera-tive	1-20,000	100	120	Saturated	95	-40 to -100	None	Electrical +7% Purge Air, Steam + 7% Purge Air, or Dry Air (15 to 35% of System Capacity)			Required Recommended	Indoor & Outdoor

1. Some deliquescent dryers have built-in afterfilters. Do not add an additional energy user to the system if unnecessary.
2. Higher flow rates will not damage, but air quality is reduced and pressure drop is increased. Not sensitive to oil and particulate.
3. The thermal refrigeration type of refrigeration dryer is the only one that does not run continuously. A thermostatically controlled switch turns the refrigeration unit on as needed.
4. Coalescing required with lubricated compressors, non-coalescing with dry compressors.

Figure 4-16   Installation Arrangements for Compressed Air Dryers
Compressed Air and Gas Institute

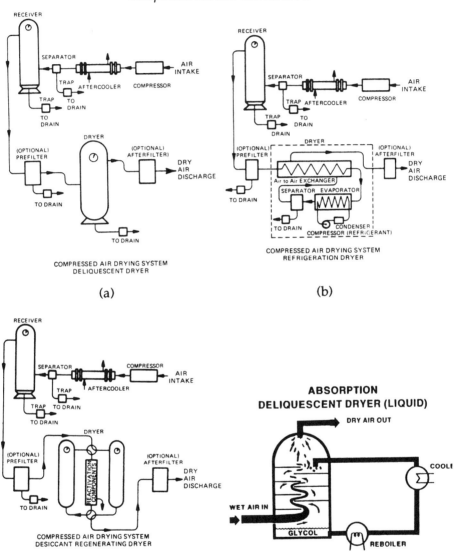

Figure 4-17   Dewpoint Conversion
Compressed Air and Gas Institute

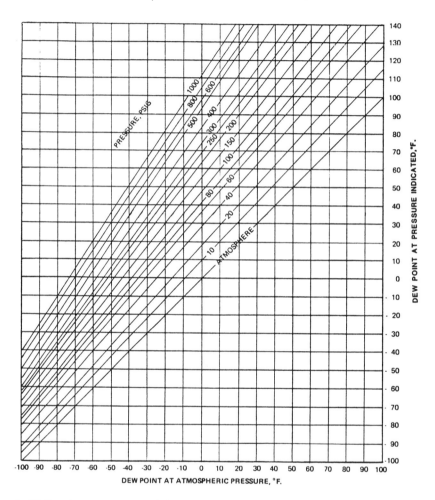

Dew Point Conversion:

To obtain the dew point temperature expected if the gas were expanded to a lower pressure proceed as follows: —

1. Using "dew point at pressure," locate this temperature on scale at right hand side of chart.

2. Read horizontally to intersection of curve corresponding to the operating pressure at which the gas was dried.

3. From that point read vertically downward to curve corresponding to the expanded lower pressure.

4. From that point read horizontally to scale on right hand side of chart to obtain dew point temperature at the expanded lower pressure.

5. If dew point temperatures of atmospheric pressure are desired, after step 2, above read vertically downward to scale at bottom of chart which gives "Dew Point at Atmospheric Pressure."

Some general guidelines for cost effective dryer selection are:

- The level of drying performance provided by regenerative dryers may not be required by a manufacturing plant air system. These dryers also require filters at the inlet and discharge which contribute several percentage points of energy loss to the system. In most cases deliquescent or refrigeration dryers are adequate, unless outside air lines require cold weather protection. Recent developments have greatly reduced the power requirements of this class of dryers.

- The pressure dew point produced by deliquescent dryers is too high for most factories in moderate to cold climates. Even with an 85°F aftercooler discharge temperature, the 100 psig pressure dew point would be 65°F. However, the mechanical simplicity, freedom from an external power source, and low purchase cost make them attractive for large intermittent flows where a low pressure dew point is not required. Although not externally powered, the drying medium in these units must be replaced and afterfilters must be used to prevent the corrosive medium from entering the system. These factors contribute to the energy cost of the dryer. Recently, liquid desiccant dryers with automatic desiccant regeneration have been introduced. They may be more convenient and economic.

- Refrigeration dryers can accept inlet air at temperatures up to 130°F and produce pressure dew points as low as 35°F. These dryers operate at moderate cost (between that of deliquescent and regenerative dryers) and produce air quality suitable for most indoor plant operations. For these reasons they have become the most popular type of compressed air dryer.

Compressed Air Filters—Filters in the compressed air line remove oil, water, and solid particles for protection of downstream components. Proper sizing of filters is critical. They are designed for specific flow capacities and pressures, and these limits should not be exceeded. Selection should not be made on the basis of pipe size. This is a common mistake in many compressed air systems.

Operating temperatures for most compressed air filters are limited to 120°F in the presence of liquids, but some may be used to 250°F when filtering solids only. Models with polycarbonate bowls are limited to 120°F and are not recommended for air supplied by compressors using synthetic lubricants.

Table 4-15 presents specific application and performance information for the three primary types of compressed air filters. Manufacturing plant air systems usually use adsorbing filters only to prevent oil contamination of regenerative dryers. Therefore, our primary concern here is the standard mechanical air filter and the coalescing filter.

Standard mechanical filters, when used in conjunction with aftercoolers, separators and receivers, can remove enough solids and aerosols for most factory air systems. Additional filtration, if required, can be provided at the point of use. This is more efficient than adding an additional pressure drop in the primary air supply. These filters suffer a 3 to 6 psi pressure drop at rated flow when removing liquids. The actual drop is dependent on incoming air quality. Design filtration efficiency is provided if replacement is performed when the pressure drop across the filter is 10 psid. However, maximum energy efficiency is obtained if more frequent replacement is performed.

Coalescing filters are used to provide air with low oil content in lieu of installing non-lubricated compressors. The system specification should consider the oil content of the air very carefully and specify coalescing filters only if required. Here again, it is best to apply these filters at the point of use if there are just a few isolated applications for oil free air.

Table 4-15　Compressed Air Filter Performance
Varigas Research, Inc.

Type	Application	Max Flow (scfm)	Pressure Drop @ Max Flow (psid)		
			Initial (Dry)	Operating[1] (Wet)	Replacement[2]
Standard Mechanical	Liquid dispersion aerosols & solid particulate to one micron	44,000	1	3-6	10
Coalescing	99+% of Condensation aerosols and 100% of particles of 0.025 micron	16,000	1	3-6	10
Adsorbent	Oil Vapor[3]	16,000	1	1	1*

1. Varies with quality of inlet air.
2. Recommended for maximum filtration efficiency. Maximum energy efficiency requires more frequent replacement.
3. Additional treatment required for breathing quality air.
*Pressure drop remains constant but cartridge requires periodic replacement.

In some applications for removing large solid particles, it is desirable to install a standard filter in series with the coalescing filter used for removing oil and water aerosols. But there is an energy cost to such double filtration.

When planning filters at work stations, the additional 2 to 6 psid contributed by the filter at rated flow should be factored into equipment selection and distribution system sizing.

Regulators and Lubricators—Regulators and lubricators have been combined with point-of-use filters to provide local conditioning of air for tools and other lubricated devices. Too often, combinations of these elements are used when only one is required. If the distribution system is properly designed, regulators should be necessary in only a few locations where system pressure is too high for the equipment in use. Some pneumatic tools and other devices may have a maximum safe operating air pressure below that available from the compressors. Care should should be exercised to avoid exceeding such pressures at all times. As described in Section 3.11, the use of regulators can reduce the quantity of air required to operate air cylinders. The addition of a lubricator adds a 5 psi pressure drop to the distribution line. Lubricators should be used only when specified by the manufacturer of the operating equipment. When they are necessary, the additional pressure drop should be considered when selecting the operating equipment and when designing the distribution system.

Traps and Drains—Drains and traps function to collect the contaminants removed from the compressed air system. Expulsion of these contaminants is critical to proper system performance. There are two primary methods for removing condensed contaminants: manual valves and automatic condensate traps.

Manual draining is the simplest and surest method of removing condensate from the system. However, this technique is not practically reliable and is useful only for small volume applications where condensate buildup is slow. When properly used, this method removes all condensate while minimizing the loss of compressed air because the drain valve is secured tightly by the operator. However, manual valves often are left "cracked open" to provide constant draining of the condensate. Such practice can be a major source of compressed air leakage.

Automatic condensate traps are used in high flow volume situations where removal of condensate must be virtually continuous or exist in a variety of configurations which have evolved to provide proper drainage of oil and water while minimizing the compressed air loss. Two main problems persist with automatic traps and drains:

• assurance that liquid is being drained from the system

• fouling of mechanical components, preventing proper drainage or positive sealing. In either case these are system problems because traps that are fouled closed permit condensate entrainment back into the system and fouled-open traps are a major source of air leakage.

Motorized drain valves with automatic controls for sequencing have become a popular technique for solving condensate draining problems. This technique incorporates some of the better features of manual draining into automatic traps, but sometimes it may be difficult to match the drain cycle with the accumulation of condensate. Motorized valves permit larger drain openings (more diffcult to clog) and positive valve closure, while the automatic controls provide the proper drain cycle.

Regardless of the type of drain used in compressed air systems, proper installation of these devices is of critical importance. The preferred method for installing condensate traps is shown in Figure 4-18.

Silencers—Air inlet silencers can be a source of pressure drop on the compressor inlet and should be used only when required by local conditions. Quantitative information is provided with the discussion of inlet filters.

Silencers usually are also provided with blowdown valves and other high velocity discharges. These silencers have no impact on system energy consumption, unless seriously clogged, but they can reduce compressor capacity especially if incorrectly sized or inadequately maintained.

Instrumentation—Strategic location of pressure and temperature gauges throughout the compressed air system can provide speedy identification of system problems and can signal the need for system maintenance.

Temperature gauges are recommended at both the aftercooler and dryer discharge. They can be useful also for monitoring the cooling water inlet and discharge temperatures.

Pressure gauges in the header and drop piping can be used to ensure that pressure drop through the system is not excessive. Differential pressure gauges should be installed across all air conditioning equipment to ensure that the flow passage is not blocked. Pressure differential is especially useful for monitoring filters which constantly are gathering contaminants. Also it can signal the build up of fouling deposits in both aftercoolers and dryers. Currently there are automatic monitors available which can provide an electrical signal when a preset pressure drop has been exceeded.

4.2.5 Selection of Compressors, Drivers and Controls

When the plant air maximum requirements have been determined, and the distribution system designed, a specification can be written for the compressed air production portion of the system. This will include one or more compressors, drive motors, compressor controls, drive motor controls and the air conditioning equipment. Most of the available options for all of these items have been described in earlier parts of this book.

The selection of specific compressor, driver and control mechanisms must consider a number of interrelated variables.

Figure 4-18
Varigas Research, Inc.

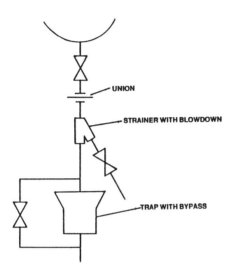

UNION

STRAINER WITH BLOWDOWN

TRAP WITH BYPASS

The specification of any one of the three influences the other two. Some factors involved in this selection are:

(1) Maximum air flow capacity requirement of system (scfm).

(2) Pressure required into air conditioning equipment.

(3) Air quality required.

(4) The energy sources available in quantity sufficient to meet the flow requirement.

(5) Feasibility of employing heat recovery.

(6) Load profile of the system—air flow as percent of maximum flow as function of time for typical 24 hour day, including any seasonal variations.

(7) Physical space available for mounting compressors—space, location, foundation, noise, and ventilation restrictions.

(8) Availability and cost of cooling water.

(9) Cost of power.

(10) Operating energy efficiencies of various compressor types under all load conditions.

(11) Estimated maintenance costs.

(12) Existing compressor types in old system.

An initial recommendation for compressor type, number of compressors and control mechanisms is presented in Table 4-16. These recommendations are to be considered as a starting point only, and they only consider factors 1 and 6. The final decision must include consideration of each of the other factors.

Table 4-16 Initial Recommendation of Compressor Number, Type, and Control Mechanism for Various Combinations of Maximum Flow and Duty Cycle Range

Max Flow (cfm)		85-100%	70-100%	50-100%	25-100%	0-100%
<100 (<25hp)	a.	One	One	One	One	One
	b.	Rec	Rec	Rec	Rec	Rec
	c.	A	A,B	B	B	B
100-700 (25-140hp)	a.	One[1]	One[1]	Three	Three	Three
	b.	Rec or Rot[3]	Rec or Rot[3]	Rec or Rot[3]	Any	Any
	c.	B, C or D	B, C or D	F	C, D, E, F	C, D or F
700-1500 (140-300hp)	a.	One[1]	One[1]	Three	Three	Three
	b.	Rec or Rot[3]	Rec or Rot[3]	Rec or Rot[3]	Rec or Rot[3]	Rec or Rot[3]
	c.	B, C or D	B, C, or D	Any	F	C, D or F

1500-3000 (300-600hp)	a.	One[1]	One[1]	Three	Four	Five
	b.	Any[3]	Any[3]	Any[3]	Any[3]	Any[3]
	c.	Any	Any	Any	F	F
3000-6400 (600-1250hp)	a.	One[1]	Two[1]	Three	Four	Five
	b.	Rec or Cen	Cen	Cen or Rec	Any[3]	Any[3]
	c.	C or E	D, E	C, D, E or F	F	F
6400-30,000 (1250-6000hp)	a.	One[1]	Two[1]	Three	Four	Five
	b.	Cen	Cen	Cen or Rec	Cen or Rec	Any[3]
	c.	D, E	D, E	C, D or F	F	F

Key

Line a = Number of compressors, b = Compressor Type, c = Control Type

Compressor Type: Rec = Reciprocating, Rot = Rotary PD, Cen = Centrifugal

Control Type: A = Unloading   C = 5 Step   E = Inlet Guide Vanes
                  B = 3 Step   D = Inlet Throttling   F = Sequential

Notes:

1. If in-house standby compressor is required, three compressors are recommended (each 50% FL) perhaps employing sequential controls

2. Sizing of multiple compressors   3 Units 50% FL each   4 Units 30% FL each   5 Units 25% FL each

3. Heat recovery feasibility favors rotary oil flood PD compressor.

The effects of these and some of the other factors are outlined below:

(1) Maximum air flow requirement of the system—This is known from the system specifications. However, the cost of system down time for unscheduled repairs must be considered. The cost must be compared with that of providing in-house standby compressor capacity. This standby capacity can be provided by two 100 percent capacity compressors or three 50 percent capacity units. The use of three compressors is desirable for widely variable load profiles if multiple compressor sequencing control systems are used. These controls automatically reduce energy consumption by operating the minimum number of compressors at any time and keeping all but one unit operating at full load. One compressor is operated in a modulation or multistep mode to handle small variations in system demand. Other compressors are turned on or off as needed to meet the system demand.

(2) Air Pressure at Compressor Discharge—The system layout, distribution system design, and air conditioning equipment specification will determine the air pressure required into the conditioning equipment. This air pressure must be provided by the compressor at the aftercooler discharge.

(3) Air quality required—a requirement for oil-free high purity air normally indicates the use of nonlubricated compressors. Such air also may be provided by lubricated compressors, but high quality oil coalescing filters must follow the compressors in the air stream.

(4) Energy sources available—The majority of air compressors operate on three phase electric power. If high pressure

steam is available or can be made available at reduced cost, the use of a steam turbine may be desirable. The turbine offers variable speed control and will serve well as a driver for centrifugal or, geared down, for double acting reciprocating compressors.

(5) Feasibility of employing heat recovery—If a large quantity of low grade variable supply heat can be employed for supplemental space or process heating, the use of a heat recovery system can become the highest priority factor in the selection of an air compressor. In smaller size compressors (below 700 cfm), the oil flooded rotary screw compressor is most amenable to heat recovery systems. In larger sizes, the rotary screw maintains a significant heat recovery efficiency advantage due to its single stage, higher temperature heat exchanger capability. The rotary screw does have a lower specific air compression efficiency, but recall that 80 percent of the input energy is available for heat recovery while only 20 percent remains in the air stream.

(6) Load profile of the system—The variation of demand for air with time is identified as a major factor in Table 4-16 It is important to consider low consumption (20 percent full load or less) periods of any significant duration such as lunch breaks, shift startup, second, third, and between shift operations. These low demand periods are typically very inefficient operating periods for single compressor systems with limited control ranges.

(7) Physical space available for mounting compressors—The practical aspects of where the compressor will be mounted, the cubic feet of space available, the strength and footings of the foundations, the noise level that can be permitted, and the ventilation restrictions that may exist are important in compressor selection. The rotary and centrifugal units

produce less vibration and, hence, can use lighter foundations. They normally require less space, especially since a receiver is not as necessary. Some rotary and centrifugal compressors are available in a noise abating enclosure to meet specific noise level requirements. A limited availability of ventilation may require the use of water cooling which favors the reciprocating and centrifugal compressors.

(8) Availability and cost of cooling water—Cooling water and sewage costs have increased several fold in recent years for many communities. A price of $0.25/1000 gallons is fairly common.. These high costs may favor the use of air cooled aftercoolers and compressors. This is especially true if heat recovery is proposed.

(9) Cost of power—The cost of power to drive the compressor sets the baseline against which the cost of the other features, such as control mechanisms, cooling methods, and driver type, are judged. Areas with unusually high or low power costs can justify different choices on each of these items. The final selection of each feature should be based on a total cost comparison, but power cost is the most important ingredient since it is the biggest single element in the lifetime system cost.

(10) Operating efficiencies of various compressor types—The typical operating efficiencies of each compressor type for an assumed load profile was presented in Figure 4-3 . The specific full load efficiency of each type was presented in Figures 4-1 and 4-2. These relative efficiencies must be considered in any total cost comparison of different compressor types. The user cannot assume all compressors require 20-25 BHP/100 CFM, as some others have recommended. The efficiency at part load is especially important for one and two compressor systems.

(11) <u>Estimated maintenance costs</u>—Most statistics do not differentiate between compressor types when presenting maintenance costs, but there are differences. Manufacturers should be interrogated, other users consulted and specific past experience employed.

(12) <u>Existing compressor types in old systems</u>—In some circumstances, it may be desirable to purchase the same type of compressors currently in place for maintenance economy, personnel training and possibly spare parts. However, this factor is minor compared to the other cost ingredients.

If several large, widely distributed air consumers are included in the system, it may be desirable to use multiple compressors with one located near each major consumer. A multiple compressor control system may be desirable, depending upon the load profile of the other, smaller air consumers. Such installations are not common, largely because of control complications and the tendency to find more than one compressor operating at part load most of the time. This is uneconomical. However, when multiple compressor locations are indicated, each compressor must be selected as recommended above.

4.2.6 Energy Saving Features

The suggestions provided to this point have been directed toward the measures to be taken in design, equipment selection, construction, and installation of compressed air systems so that they may deliver the amount and quality of air needed to support the using operation at the least expenditure of energy. There are some compressed air system equipment items offered by an imaginative and responsive industry which are directed specifically to the purpose of saving energy, and these are worth noting.

One of these is the reheater which is installed immediately following an aftercooler or dryer. The air-to-air reheater following a refrigeration drying cycle is well known, but recently some companies have offered oil heated reheaters. These recover heat from the cooling oil of rotary screw compressors to increase the temperature of the discharge air after aftercooling and/or drying. This further reduces the relative humidity and provides a larger volume of air to the users. If this higher temperature air is not objectionable (on the order of 165°F), then such a reheater should be seriously considered. It will save money, but may reduce dynamic valve seal life, accelerate deterioration of hoses and valves, and require insulation of drop lines and equipment.

Another potential conservation measure, of course, is the possibility of recovering heat through the heat recovery measures described in Section 4.1.5. Most of the air compressor manufacturers offer heat recovery packages for their air-cooled compressors exceeding 50 horsepower and water-cooled ones exceeding 125 horsepower. Other manufacturers offer some unique modules which incorporate compression and heating into one acoustic enclosure to facilitate distribution of filtered hot air to the factory floor.

In evaluating the applicability of the heat recovery packages, the total annual value of the recovered heat should be considered, mindful that the heat is not likely to be useful in summer for space heating. In certain processes and operations, this heat might be useful for other purposes.

In considering the economics of such measures, an allowance should be made for the efficiency of alternate heat sources. For example, most gas fired overhead unit heaters in factories are not particularly efficient and when their part load factor is considered, they will, over a period of time, require about 1.6 Btu of potential fuel heat content to produce 1 Btu of effective heating. The problem is that each time the heater turns off, it cools, discharging most internal heat through the vent pipe to the outdoors. Some other types of factory heating may not have so bad a performance, but a realistic estimate of the heating efficiency should be made and factored into the analysis.

Compressor manufacturers are offering new compressor designs which target lower power consumption. These usually relate to controllability features such as inlet throttling, sump blowdown and multi-step unloading. Sometimes these features increase the cost of the compressor, but in most cases, they can be cost justified through power saved. They should be considered seriously and, in most cases, probably should be purchased unless the compressor is expected to be operating either very infrequently or at full capacity all of the time.

Recent developments are offering other distribution system devices such as the flow sensor which shuts down a line which has developed an exorbitant leak and a new type of line sensor which senses flow condition and shuts the line down when the flow is at leakage rather than work load rates.

System maintainability is important. Everything that may leak or otherwise need service should be accessible. Difficult or vexing maintenance tasks tend to be postponed. Potential sources of leaks are more likely to be checked if they can be seen and reached.

Some compressor manufacturers are offering total compressed air energy management systems, others new microprocessor controls for multicompressor systems. Before any compressors are purchased and systems designed, the major equipment vendors should be interrogated about the latest available technology. These, then, should be cost analyzed as recommended in section 5.2.

One of the most important aids to energy conservation in distribution systems, however, is simply to make sure that they are well instrumented. Pressure gauges, pressure differential gauges, and thermometers can facilitate system and equipment troubleshooting. This is discussed in greater detail in Sections 4.2.7 and 4.3. Good instrumentation will be found to be necessary for effective maintenance supervision.

4.2.7 Institution of Maintenance Program

As a part of every new or modified installation, the maintenance program should be organized and initiated, or reviewed if a program already is in place. It is important that maintenance personnel participate in the system startup and first leak detection. New systems almost always have some leaks and some debris in the line which causes maintenance troubles in traps and filters in the beginning. Filters and strainers will need to be checked more often during the first few weeks of operation than later.

Another measure that should be taken is the determination of base line readings of all pressures and temperatures when the system is operating normally. Also, the compressor pump-up time and the system decay time should be measured when new, in good condition and leak free. The recommendations in Sections 4.3 and 5.1 should be followed.

Starting a first class maintenance program coincident with beginning to operate the system should be considered just as important a matter as the selection of any of the major pieces of equipment.

One person should be responsible for compressed air system maintenance, but the maintenance, engineering and production departments should be involved. Follow-up should be provided by the plant general management. Some companies use multipart tags for maintenance items with copies going to several departments. This maximizes the chances of follow-up.

4.3 Maintenance

There are few industrial systems for which maintenance can have so profound an effect on energy consumption as it can with compressed air. There are many areas of potential inefficiency, and maintenance inadequacies can show up as increased energy consumption in several ways:

    decreased compression efficiency
    leakage of compressed air
    excessive pressure drop
    reduced pressures at user stations

Inadequate maintenance also can affect efficiency indirectly:
    inadequate control of temperature
    inadequate control of moisture
    excessive contamination of air

All seven of these problem areas can have secondary effects on other pneumatic equipment down the line.

All of the equipment in the compressed air system should be maintained in accordance with the manufacturer's instructions, and detailed recommendations will not be repeated here. However, there are maintenance requirements to which energy use is especially sensitive, and the ways that these maintenance failures can affect system efficiency will be discussed, with recommendations.

4.3.1 Compressor Drives

Failure to maintain an electric motor properly is more likely to result in eventual failure than in a mere increase in power consumption, but there is no doubt that a failing bearing does contribute added friction, and obstructed motor cooling will increase motor temperature, winding resistance, and hence, electrical losses. Also, improper maintenance of the motor controls can reduce motor voltage, causing an efficiency reduction.

Another area worthy of attention is that of V-belt drives. Belts which are too loose will slip and those which are too tight will excessively load bearings. Wearing and stretching of belts in normal use requires that they be periodically inspected and adjusted. Belt driven compressor installations should be monitored regularly for proper belt tension. They should be checked at intervals not exceeding 500 operating hours. Worn or frayed belts should, of course, be replaced.

4.3.2 Compressors

Compressor maintenance has a severe impact upon system energy efficiency. The important maintenance failings of compressors are:

wear of mechanical parts
fouling of cooling surfaces
dirty air filters
inadequate or dirty lubricants

No great insights are required to predict that when compressor metal parts performing a sealing function of some sort fail, then the compressed air will short circuit internally from high pressure areas to low pressure areas. Such parts in reciprocating compressors subject to wear are piston rings, valves, and cylinder walls. One of the virtues of lubricated rotary screw compressors, and possibly the major reason they have become popular, is that the nature of their construction significantly decreases the potential for wearing of metal parts. There simply are not as many metal parts coming together, and those that do are fully lubricated. However, all compressors should be evaluated periodically to determine that they are able to compress to full capacity and that there is no increase in the ratio of power to output.

Another area of vulnerability arises from the thermodynamic effects of fouled compressor and intercooler cooling surfaces. When this occurs, the increased temperature rise in the compression process has the effect of reducing the output in scfm, but not the total energy input. This added energy per unit output shows up as higher temperature of the air discharge from the system. This temperature normally then is reduced by the aftercooler, the dryer and the distribution system. The final result is less compressed air power to meet factory demand. It is important then that all cooling surfaces and cooling media handling equipment such as fans, and water pumps, be maintained to top performance.

A dirty inlet air filter can cause a reduction in capacity but only a negligible loss in efficiency. If an inlet filter is left in service until the pressure drop increases by 3.5 inches or 1 percent of atmospheric pressure, then the capacity of the compressor will have been reduced by 1 percent.

Because of the destructive effects of dirt particles in compressors, especially the rotary types, it is crucial that the filtration be kept effective and all inlet piping be kept clean. The user should select and install good filters and maintain them regularly. The maintenance interval depends upon the nature of the environment from which the suction air is drawn.

Lubricated compressors depend upon oil for both friction reduction and sealing. It is obvious, then, that permitting the oil to become contaminated or corrosive is destructive of both efficiency and valuable capital equipment. The oil quality is especially important to many rotary positive displacement compressors because the entire compression process is oil bathed and oil cooled.

In bringing about proper and diligent maintenance of compressors, there is no substitute for regular inspection and testing. Some tests that can be applied to evaluating compressor and compressor accessory condition will be set forth in Section 5.1.

4.3.3 Compressed Air Conditioning Equipment

When the air leaves the compressor, it may enter any one or several pieces of compressed air conditioning equipment as described in Section 3. Each of these is an area of potential maintenance failure. Compressed air filters are among the highest pressure drop items in the system, even when clean. When they have become fouled, the loss potential is obvious. An added 1 psi drop in a 100 psi system represents 1 percent loss in energy availability to the air user. The filters must be kept clean.

The same can be said for the aftercoolers, dryers, separators and any other equipment in series with the air flow. A system that is equipped to deliver moisture-free air, or even dry air, should be maintained to ensure that the desired product is produced. In this way corrosion and scaling of pipes will be reduced, along with the incident pressure drop.

4.3.4 Traps and Drains

Moisture traps are located in many points throughout the system. They normally are at receivers, separators, filters, coolers, and drying equipment and in low points throughout the distribution system. Since these traps are mechanical devices with moving parts, they can fail either open or closed. If they fail in the open position, they are a very costly leak. If they fail in a closed position, the result is liquid buildup in the system, with several potentials for reduced performance.

In many air systems the traps are much neglected. Sometimes they are bypassed or replaced with valves left in a partially open position. Not only is this wasteful of air, but these valves can clog and cause liquid entrainment in the compressed air. Properly installed traps, with protecting strainers should perform well for many years; however, they require periodic inspection and service.

4.3.5 Filter/Lubricators

The filter, regulator and lubricator are used to cleanse the air at the point of use, to regulate the pressure and thus the power or thrust of the tool or other pneumatic equipment, and finally, to lubricate that equipment, in that order. If the tool or other pneumatic equipment is not protected (by a filter) from serious contaminants, or if the equipment is not properly lubricated, it may wear more rapidly and thug may reduce efficiency and expend more air to accomplish the game job.

Any clogged filter will have the effect of added pressure drop and a resulting loss of energy as explained previously. Only regular inspection and attention can keep these items in proper order.

The in-line lubricator is available in several designs. Different air users may require different types of lubricators. Manufacturers' tests have indicated that proper lubrication of air tools results in reduced air consumption for governed tools (up to 50 percent compared to dry tools) and increased speeds for ungoverned tools.

4.3.6 System Leakage

Of all of the maintenance failures, system leakage probably results in more lost compressed air energy than any other single factor. Plants have been observed where leakage losses are a modest 10 percent of the total compressed air capacity. Although this is "modest" by leakage standards, it is a significant annual dollar cost. Other plants have been observed with leakage rates in the range of 20 to 40 percent of total air usage. The cost of this leakage is high, avoidable, and reprehensible.

The table below shows a conservative estimate of the annual cost of leaks of various sizes:

Equivalent Hole Diameter	Leakage Rate scfm	10^3 scf per year (4000 hrs)	Cost per year, $ (40¢ /1000 cf)
1/64"	0.25	60	24
1/32"	0.99	238	95
1/16"	3.96	950	380
1/8 "	15.86	3,806	1,522
1/4 "	63.44	15,226	6,090
3/8 "	142.74	34,258	13,703

Air at 100 psig
Orifice with sharp edges (Coefficient of flow = 0.61. Leakage and cost could be increased 60 percent for well rounded hole coefficient = 0.97).

As can be seen, several small leaks go a long way. The sources of these leaks have been found to include any device or connection in the system, especially:

threaded pipe joints
flange connections
valve stems
traps and drains
filters

hoses
connectors
operating valves on pneumatic devices
check valves (back flow leakage)
relief valves

Sometimes leakage is induced by operators who "need a little air" for unauthorized purposes such as self cooling.

The fight against leaks is never ending and must be considered such by any diligent maintenance department. The work is never done in a productive and active factory because changes always are being made. Sometimes there are production increases, requiring new drops and hoses, or reductions, requiring that some drops be abandoned, but perhaps still leaking, or relocations, requiring both additions and deletions. Valves, traps, and connectors wear and pick up flakes of contaminant and will commence leaking on their own initiative. The inspections and searches for leaks must be regular, diligent, and thorough. Some suggestions for determining the total amount of air being leaked and seeking out specific leaks are included in section 5.1.

4.3.7 Excessive Pressure Drop

If all of the above measures have been taken to reduce losses and maximize the efficiency of the distribution system, but the pressure drop still is higher than the design calculations predict, then the piping system should be examined for internal blockage or corrosion. As distribution lines age, corrosion and contamination collect in low areas and at joints. Sometimes these accumulations can migrate and reaccumulate, forming a larger blockage at a particular location.

4.3.8 Cumulative Effects

The cumulative effects of leakage, pressure drop and compressor under-performance can combine to produce effects which are thoroughly puzzling.

It is important in the management of compressed air systems to realize that compressed air is an expandable medium, and when, through excessive pressure drop and/or excessive leakage, it is forced to be distributed to its end use at too low a pressure, then its energy content is vastly reduced. The inevitable effect is a combination of reduced factory productivity and greater air usage, even further reducing pressure and efficiency.

For example, when there is too much leakage, the air flow necessary to operate the factory is augmented by the air flow necessary to accommodate the leaks. This can cause a higher pressure drop in the headers, reducing the pressure and, hence, the energy content of each cubic foot of air. This, in turn, increases the amount of air that the factory needs to do its job, which, in turn, further increases the amount of air drawn from the headers, expanding it, further reducing its energy content. Some factories have experienced work station pressures of only 40 percent of riser pressures. There have been cases where the addition of compressors simply could not increase work station pressure because the system had "self-blocked" with this low pressure, low energy content air.

Even adequately sized systems have been known to "self block" because excessive leakage dropped pressures to such a low point that the total energy transmission capacity of the system became inadequate. The solution is to shut down all air use long enough to restore the pressure (and air energy content) again before reloading the system. A 2- to 5-minute shutdown usually is adequate to bring system energy delivery capability up. The system deficiencies (usually excessive leakage) that caused the shutdown should be corrected as soon as possible. A well designed and well maintained system will not experience such events.

The economics of all of this indicate that it is almost impossible to overstate the need for good maintenance.

4.4 Operations

There are a number of opportunities for conservation in the methods by which compressed air systems are operated. The operational improvement suggestions which will follow are not meant to be a final and complete list. There are others which experienced managers of compressed air systems can offer as well. Some of the desired operating practices discussed below may not be possible if the pressure controls and some other features of the system are not so designed and constructed as to offer the full range of flexibility desired. Under these circumstances, some minor system modifications might be indicated so as to have the system more closely resemble the design and construction recommendations made elsewhere in this report.

4.4.1 System Operating Schedule

One of the most energy costly practices in industry is that of providing available compressed air pressure during times when operations are not taking place. One single shift factory was observed to be operating compressors and maintaining full system pressure seven days per week, twenty-four hours per day as a carryover from practice during busier times. There are a few small sets of instruments and other operations which do need full time air pressure for their proper functioning, but this requirement is miniscule by comparison with the cost of operating a large compressor throughout the weekend. It is obvious that a small compressor to handle these vital functions and a shutdown of the rest of the system will be immediately cost effective.

The same can be said for overnight operations. Some factories operate the compressors for similar reasons all night, even though there is no third shift. Once again, a small compressor can do the job for the few vital continuous functions.

The schedule also should be examined for other available shutdown periods. For example, at what time does the compressor really need to be started in the morning and how soon can it be shut down in the evening? In smaller factories, is it possible that noon-time shutdowns would be economical?

Where multiple compressors are in operation, the system should be scrutinized to assure that the most economic array of compressors is on stream at all times. For example, a night shift operation may have only one half of the maximum air requirement as the day shift operation; therefore, some of the compressors can be pulled off-line for that shift.

There may be other opportunities to save energy through proper air compressor management and scheduling.

4 4 2 Compressors and Controls

The air system should not be operated at any higher pressure than that required to deliver the design pressure at the points of use. Operating tools at higher than design pressure increases the maintenance costs and may be unsafe for the operators, and reducing a higher line pressure to the design pressure is an energy and cost loss. Therefore, the compressor control system should be set to maintain the optimum pressure. The practice of generating to a high pressure for buffer storage and reduction for use is wasteful. Some multiple compressor control systems will automatically shut down some of the compressors when the demand for air has fallen. Such controls should be reviewed periodically to assure that their programming is up to date with changing factory operations. If more buffer storage is needed, add another receiver.

4.4.3 Dryer Operations

Once again, a scrutiny of schedules should be made to determine if fully dried air is needed at all times. Some systems are equipped with two dryers in series, a deliquescent one for keeping the distribution system generally corrosion free and another more effective dryer for portions of the system where moisture is more critical. It is possible that during certain shifts the second dryer is not required and can be bypassed.

Energy economy almost always can be effected by improving the control of the timing of the purge cycles of desiccant dryers. These cycles should be geared to the actual air consumption rather than continuous automatic operation. This can be done by a rather simple timing or flow measuring system. An alternative is to control the timing of the purge cycles of the dryers on the basis of outside air humidity, or, even better, desiccant moisture content. Automatic controls are available for such operations with savings of 75 percent of the power previously used on the dryers having been reported. Similarly, the operating cycles of refrigeration dryers should be controlled based on the flow of compressed air through the system.

4.4.4 Use of Compressed Air Devices

There is a host of energy saving opportunities in the final application of compressed air. First and foremost, can a more energy efficient utility than compressed air be applied to the task? (See Bibliography, Item 15.) Productivity considerations may require air tools, as labor costs more than power and pneumatic tools are productive. But, in one case, some large stationary routers were being driven by air motors consuming 320 scfm each. An electric conversion was scheduled for a saving of more than 80 percent of the energy going into pneumatic operation of those routers. In this particular case, there was no effect on labor costs.

When compressed air is clearly indicated for productivity or other reasons, then it should be used as efficiently as possible. For example, in some punch presses and automatic cut-off tools, air is used to discharge parts. Rather than have the air blow continuously at the part being cut, it can be intermittently applied when needed through the use of suitable control valving. In the same type of installation, it has been found that a small laminar flow nozzle consumed much less air than a pinched piece of tubing.

When applying compressed air, it is well to keep in mind the cost of air leaks as described in Section 4.3.6.

4.4.5 Partial System Shutdown

Sometimes factory changes result in temporary or semipermanent abandonment of a section of the system. Under these conditions, the appropriate headers should be closed off either by a valve, if it exists, or by inserting a blind flange, if possible.

The same can be said for individual machines and tools. The valves used for tool actuation are operated repetitively and tend to wear out and become leaky. Therefore, additional manual valves should be closed whenever a work station or a tool is not in use. This is true especially of the large stationary tools which tend to have a number of pneumatic functions built into them. If operator shutdown of such unused lines is not reliable, then management should install solenoid valves which are operated by machine electrical power or lighting.

4.4.6 Operational Inspections

The normal operation of the compressed air system should include the assignment of a designated person to inspect the system regularly while operating at full load. This inspection should be performed daily or at least weekly. Items that require less frequent inspection can be scheduled on a rotating basis.

The differential pressure gauges should be read to determine pressure drops across all components of the air conditioning system, especially the dryers and filters. These should be monitored daily, or at least weekly, as part of normal operations. The maximum differential pressure allowed before cleaning or changing a filter should be indicated on or near the gauge so that corrective action can be initiated when indicated.

The operation of all traps and drains should be checked. The trap bypass should be operated periodically to assure that the condensate is being regularly drained. The trap should be flushed occasionally. Any readily noticeable leaks should be noted for necessary maintenance work.

4.4.7 Alternative Systems

Previous Department of Energy studies (one of them reported in the Bibliography, Item 15) have determined that the practice of delivering energy to manufacturing operations by compressing and then expanding air is a relatively inefficient process. The compression of air creates a good deal more heat than potential pneumatic energy, and then the delivery of this energy through piping systems, with their losses, further reduces the power available at the tools and motors. When this air, then, is re-expanded in the using device, it is, in most cases, not fully expanded for reasons of temperature and tool or motor size.

The result is that even in an ideal and superbly maintained and operated compressed air system, the ultimately attainable energy efficiency is going to be very much lower than that of other systems such as direct electrically driven operations (the most efficient of all) or hydraulic systems. (See 4.4.4 above.)

However, it is understood that the manufacturers are interested in total cost, not merely the cost of energy. Compressed air systems would not be so widely used as they are were it not for their very significant contribution to the labor efficiency of production operations. The manufacturer must, of course, be motivated by this ultimate consideration, but a number of investigators have found that, many times, compressed air is applied when another system would be just as labor efficient, or at least to the extent that the ultimate total cost might be lower if a more energy efficient option were selected.

In some cases, compressed air is used because it is there. It might be the most readily available, and a user can connect with the least amount of difficulty. In other cases, it simplifies safety and maintenance considerations, although it has been found that, in most cases, electric tool installations can be made to be just as safe and properly sized ones do not present greater maintenance problems. In other cases, managements install compressed air systems because previous economic analyses, when the energy/labor equation was different, called for compressed air.

In making cost comparison studies, all factors should be included. Industrial compressed air systems not only have a much higher energy requirement, but they usually incur higher capital investment because of the need for compressors, air conditioning equipment, a distribution system and a larger electrical service.

It is recommended that these costs be reconsidered from time to time in the light of changing manufacturing operations, tool availability, labor costs, and energy costs. This is especially timely when new installations or plant expansions are being planned. Manufacturing methodology should not be determined by past practice, custom of the industry, or for any reason other than lowest, long term bottom line cost. (See Section 5.2.)

4.5 Existing System Improvement

Frequently, energy can be saved through appropriately selected modifications to existing systems. Section 4.2 above outlines a good approach to compressed air system design and equipment selection. Many existing systems in older plants do not measure up to the criteria and methods applied in Section 4.2. To this extent, they may be candidates for improvement with a resulting savings in energy and cost.

Some of the most frequent shortcomings of such systems, which increase their cost of operation and energy use, are distribution headers and branch lines that are too small, air that is too wet, blow down valves instead of traps, branch lines and hoses that are too small, connectors that are too restrictive, inadequate distribution system cutoff valving, oversized compressors, inadequate compressor controls, poor header and branch line layout, deteriorated system, undersized filters and accessories, and poor air inlet location.

It takes no great insight to understand that if a system is deficient in any of the above ways, then some corrective action should be considered. Section 5.1 offers some suggestions for system evaluation and Section 5.2 offers some suggestions for cost analyzing possible corrective action.

For example, if it has been determined that the compressor inlet air is coming from a warm area and that the pressure drop to the compressor is greater than desired, consideration might be given to moving this inlet and extending a larger inlet pipe to a cooler area. An external inlet on the shady, north side of the building is desirable. By noting the cost effect on the efficiency of the compressor, the ongoing cost of the inadequate inlet can be weighed against the cost of the modification, all in accordance with Section 5.2.1.

Sometimes the material and labor cost of an improvement, such as adding a solenoid valve or installing a larger filter, is trivial by comparison with the ongoing energy cost savings. Lists of these potential improvements should be organized so that when maintenance personnel are between crises, they can carry out the improvements on a routine basis.

One of the mistakes that frequently is made in air system management is that of noting that "the pressure is too low" so it is time to "buy another compressor." All too often, the pressure is too low because of reasons having nothing to do with the availability of air at the compressor. A distribution system may be too small, improperly routed, partially blocked, too leaky, or a combination of these. Merely adding more compressor capacity to try to force more air through such a system is not the most cost effective solution, even if it sometimes succeeds.

The value of the compressed air in a distribution system is a function of the volume that is available and the pressure at the work points in the system. This is a measure of the work delivery potential of the system. If the system is delivering air at too low a pressure, then much more air will be used to do the same job, causing a greater flow and greater system pressure drop even if the system design pressure is maintained at the compressor outlet. The solution then is to correct the distribution system shortcomings first and consider more compression capacity only if necessary.

One saving secondarily related to energy is control of cooling water. A temperature sensor to control coolant flow should be installed on all compressors and coolers in accordance with the last paragraph of Section 4.1.4. If such controls are not in place, they should be installed.

If major system modifications are being considered, the best procedure may be to follow all of the steps outlined in Section 4.2, although some of the sections may be only partly applicable. Making all modifications in the light of a total system review will yield better results.

5.0 PROCEDURES FOR SYSTEM EVALUATION

5.1 System Measurements

Basic to all decisions concerning existing systems is a knowledge of how the systems presently installed are performing. The engineering measurements of greatest interest are:

Power to compressor
Air flow from compressor
Temperature of air and coolants
System leakage
Pressure drops in system

An ideal system will have instrumentation to measure all of these things. Such systems can be closely monitored for performance and, more important, for charges in performance from new condition. Recommendations about instrumenting new systems and improved systems will be found in Section 4.

But how does one evaluate a system which is not fully instrumented?

5.1.1 Electrical Measurements

Measurement of electrical energy into the system is not difficult. There are a number of good portable metering techniques available which can do this, the easiest to use probably being hand-held wattmeters which are equipped with clip-on leads for voltage and an inductive sensor for current. These are adequate for most air compressor drive work. For installations where the compressor consumes a large percentage of the plant power, the power factor is an important consideration in operating costs and must be measured. Power factor is a major concern during part load operation. To determine power factor it will be necessary to supplement the measurement of power, in watts, with independent readings of voltage and current.

5.1.2 Air Flow

The measurement of air flow in the absence of installed flow instrumentation is more difficult. One way, of course, is to modify the system to provide a blind flange discharging to the atmosphere. An orifice of known diameter can be substituted for the flange, and the compressor operated at capacity with full output through the orifice. The flow rates of orifice diameter versus pressure drop and temperature are well known. For the most accurate orifice measurement techniques, the ASME Power Test Code for Velocity/Volume Flow Measurement should be followed.

The problems with this technique are two: first, the noise level will be extremely high. This will mandate a discharge outside of any occupied buildings, the result then being a very considerable neighborhood nuisance during the period of the test. However, the test need not be a long one. The other problem is that a modification to the piping, if such a connection does not already exist, is no simpler and in many cases is more difficult than the installation of an air flow sensor in the discharge line which could provide either continuous or occasional flow measurements. Many types are available, the direct velocity pressure probe favored for reduced pressure drop. A large orifice with pressure taps upstream and downstream can make possible the use of pressure differential instrumentation to check the flow rate from time to time, but this type of instrumentation provides ongoing pressure and energy loss at the orifice.

If the reasons for the flow measurements are to check compressor performance, quite often the case, then another approach sometimes is useful if it is done carefully, regularly, and with good and consistent record keeping. It is based upon the principle that the time required for a compressor to come to full pressure and to charge fully the receiver, air conditioning equipment, and perhaps a portion of the distribution system, is a function of the condition of the compressor. As soon as a new compressor is installed, it should be isolated temporarily from any potential leakage situations in the factory. Then simply measure the time that is required to bring the discharge pressure to a predetermined known value. The inlet air temperature also should be recorded at this time.

Later, as the compressor ages, and more users are added, the test should be repeated. If the time to charge increases noticeably, and this cannot be accounted for by leakage, then it is obvious that the compressor output has decreased. When there is insufficient output, the suction system and inlet air filter should be inspected first to make sure there are no restrictions in the suction line. Then the speed of the compressor and the unloaders and controls should be inspected to make sure they are functioning as required. For centrifugal compressors, the surge pressure should be checked, following manufacturer's recommended procedures. If all of these items are in order and there are no leaky head gaskets, then one or more loose or worn metal parts are indicated.

5.1.3 Temperature Measurements

Temperature measurements also are important indicators of equipment function. Not only will compressor cooling media temperatures indicate difficulties with the cooling system, but elevated temperature rise also can result from failing cylinder lubrication, a defective unloader or worn valves. Similarly, above normal discharge air temperature can be caused by any of the same reasons and, also, by defective or improperly adjusted output controls.

Other system temperatures are used to monitor the performance of coolers and dryers. The most significant indicator of aftercooler fouling, or obstruction, is the approach temperature.

This is the difference between the temperature of the air discharged from the aftercooler and the temperature of the coolant being supplied to the aftercooler.

All temperatures should be read when the system is new and operating properly so as to establish a data baseline. These baseline temperatures should be compared to those indicated by equipment manufacturers as being representative design or specification temperatures. This will establish the basis for interpreting future operating temperatures as the system is used.

5.1.4 Leakage Testing

Of all the maintenance measurements needed, the detection and measurement of leakage probably requires more attention from a good maintenance program than any other item. There are many methods of finding leaks.

One of the favorites for all types of leakage in most places is simply the use of a can of soapy water and a paint brush. Such a technique should be applied immediately to all new installations and to all system modifications. Every pipe joint should be checked before the installers are released from their obligations. Although the soap bubble test still is by far the most reliable detector, even of relatively small leaks, its disadvantage is that seeking leaks throughout a large system requires personnel to cover meticulously all of the joints and connections which might be suspect. This could require a great deal of time.

A more modern approach, though perhaps not quite so sensitive as the use of soap bubbles, is the application of acoustic detectors. All air leaks make high frequency hissing sounds, the larger of which are clearly audible in most factories after the work has stopped and while the compressors still are running. An acoustic leak detector can be used to detect and locate these leaks. A number of manufacturers make good ones which are portable and with which an entire system can be scanned in a reasonable amount of time. They consist of small, directional microphones, amplifiers, audio filters and either visual indicators, earphones, or both. They are effective.

An integrated evaluation of total system leakage also is desirable in many factory installations. This can be achieved by operating a compressor on a Saturday or at night when no work is being performed. If the system has air flow measuring instruments, then the leakage can be measured in this way during weekends or off hours when using operations are shut down.

If there is no air flow instrumentation, and the compressor is known to be in good condition, then sometimes leakage can be determined by measuring the output of the compressor as a percent of its total capacity by observing its control systems in operation. For example, assume that at the average of the high and low controlled pressure, the capacity of the compressor is known to be X cubic feet per minute. Also, assume that it is controlled by alternately fully unloading and fully loading the compressor. If under these circumstances the loaded time is 30 percent of the total operating time during off hours when manufacturing operations are not taking place, then the leakage can be said to be close to 0.3X.

Once again, it is wise to perform such measurements on new systems to establish baseline information for a later comparison. Also, this type of test is best done just after performing a test or measurement of compressor operating characteristics to assure that it is operating at a known capacity.

Another method has been used which combines compressor capacity and system leakage measurements in the following way:

During off hours, a compressor is started and the time to bring the system to full pressure is carefully recorded. Then the compressor is turned off or unloaded, and the time for the system pressure to decay to a known value is recorded. This first is done when the system is new and all or almost all leaks have been eliminated. The times are recorded carefully. Future tests, then, will indicate condition change. If the system pressure decays more quickly than previously after shut-down of the compressor, then an increase in leakage is indicated. If, after the leaks are brought under control so that the decay time is the same, an increase in total test time or pressure rise time indicates a compressor in need of attention.

System Leakage Test
Varigas Research, Inc.

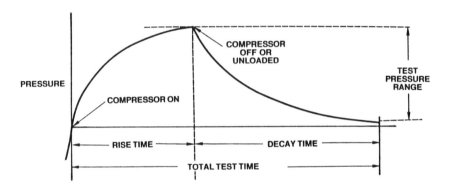

It is important to note here that, to control leakage, the responsible maintenance personnel must be in early and apply regular and continuous attention to the problem. By beginning when the equipment is new or newly refurbished, baseline information can be established and the first leaks corrected at the expense of the contractor or others who may have installed the system. Also, this will set a performance standard against which maintenance performance can be scheduled and evaluated.

5.1.5 Pressure Drop Measurements

Most compressed air equipment is, or should be, provided with input and output pressure gages. One important place where, frequently, this is not so is at work stations and places where tools may be attached, such as at the end of hoses. For this reason, every good maintenance department will have good, high-quality pressure gases for checking system pressure in remote areas. Once again, an early routine for regular pressure checks at the beginning of new system operations will provide baseline information helping to identify later difficulties with the distribution system. These difficulties could arise from overloading a portion of the system with too many drops and usages (factory operations'change), new leaks, or line obstructions. Such on-the-spot evaluations can indicate undersized branch lines, drop lines, filter/lubricators, or hoses. Sometimes higher horsepower pneumatic devices are added to existing lines, rendering them inadequate. Good maintenance operations can sense and correct this problem before the tool users become frustrated.

Reading the of actual pressure in a hose leading to a tool or portable pneumatic device is recommended. Most air motors suffer badly in operating function when air pressure is down, resulting in longer operating time and more total air used per operation. Several manufacturers offer "hypodermic" needle pressure gages which can be inserted into a hose near the end connected to the tool or motor. The gage then is read while the tool is operating.

It is important to note that in many cases the maintenance personnel will be subtracting two pressure measurements in order to determine a pressure drop, such as the drop across a filter, because some installations of these pieces of equipment do not include the preferred pressure differential gages. Since most pressure gases are not of laboratory quality and may change in calibration with time, it is important that the pressures be read at least once and, perhaps, from time to time subsequently, when there is no air flow. The no-flow pressure readings all will be of the same pressure, and a recording of the differences will quantify gage errors which then can be compensated when pressure differences are being determined. Even so, determining pressure drop in this manner is inferior to the readings from pressure differential gases.

Pressure drop measurements are of great importance for filter condition monitoring. Systems equipped with filters also should have pressure differential gages, even if later retrofitted. Pressure differential alarms also are available and effective in announcing when filters need cleaning or replacement.

5.2 Financial Evaluation

The dominant consideration in deciding whether or not to take an action that may save energy is, of course, saving money. This dictates that all comparisons of alternate courses of action, or alternate pieces of equipment, should be reduced to a meaningful financial evaluation. In the final analysis will the cost of enlarging a header system justify its cost? Will compressor A be, in the long run, less costly to purchase and operate than compressor B? Which air dryer will offer the lowest lifetime cost? Is the extra cost of high efficiency electric motors economic? When should filters be changed for lowest total cost? There are a number of financial and accounting analytical methods which will produce criteria for sound value judgements of these types.

5.2.1 Equipment and System Decisions

How do engineers and managers best decide which type of dryer, compressor or total system will achieve best economy? The lowest first cost may not result in the best buy. These decisions are important and must be made judiciously.

The approach that appears to be most useful and applicable to making system design and equipment choices is analysis based on reducing to present value all cash costs and credits, both present and future. Using this method, a rational decision can be quantified into a single cost total for each of the alternative actions being considered. These cost totals then can be compared without further qualification, if all of the in-going data are correct and the assumptions are the best and most rational that can be made.

The analysis begins by determining all of the relevant facts. What are the initial capital costs, including equipment, engineering and installation? What are the ongoing operating costs, including maintenance, energy, cooling water, cooling water disposal, and any other? How will the installation be depreciated in company income tax returns, and what tax savings are made by declaring this depreciation? What federal or state investment tax credit will apply, if any? What does company management estimate that the cost of capital is and will be over the life of the equipment? What is the salvage value?

Care should be used in developing each of these cost inputs, and the operating assumptions should be based upon actual experience and realistic future projections. Also, some reasonable projections should be made for inflation, and the inflation projections should be consistent with the assumed cost of capital. Any analysis is only as good as the information inputs.

After these cost factors have been clearly and as accurately as possible determined, the following steps should be taken:

1. Tabulate all costs by year for the life of the equipment or systems being considered. This tabulation should include the initial cost, all ongoing costs, and, as a credit, tax savings from credits and depreciation. This will yield a total cash outlay for each year of the life of the equipment.

2. Establish a discount factor to present value for each year of operation of the equipment. In order to assist, Table 5-1 is included, and it is a table of discount factors to present value for capital costs ranging from 6 percent to 26 percent for up to 15 years. Reference to this table can provide suitable factors. For capital cost values between the even numbered rates covered, interpolation is satisfactory.

3. Multiply the total cost for each year by the discount factor for the same year, yielding a present value of the cost for each year of the life of the equipment.

4. Total all of these annual present value costs for a final total system lifetime cost, discounted to the present. This is the basic figure of merit which should be used for cost comparison.

Table 5-1   PRESENT VALUE FACTOR TABLE
For Costs Incurred Throughout Each Year
(Mid-Year Present Values)

Year	Discount Rate or Cost of Capital										
	6%	8%	10%	12%	14%	16%	18%	20%	22%	24%	26%
1	0.972	0.963	0.955	0.946	0.939	0.931	0.924	0.912	0.910	0.903	0.897
2	0.917	0.892	0.868	0.845	0.823	0.803	0.783	0.764	0.745	0.728	0.712
3	0.864	0.826	0.789	0.754	0.722	0.629	0.663	0.637	0.611	0.587	0.565
4	0.816	0.764	0.717	0.674	0.634	0.596	0.562	0.530	0.501	0.474	0.448
5	0.770	0.708	0.652	0.601	0.556	0.514	0.476	0.442	0.411	0.382	0.356
6	0.726	0.655	0.592	0.537	0.487	0.443	0.404	0.368	0.337	0.308	0.282
7	0.685	0.607	0.539	0.479	0.428	0.382	0.342	0.307	0.276	0.248	0.224
8	0.646	0.562	0.490	0.428	0.375	0.329	0.290	0.256	0.226	0.200	0.178
9	0.610	0.520	0.445	0.382	0.329	0.284	0.246	0.213	0.185	0.162	0.141
10	0.575	0.482	0.405	0.341	0.289	0.245	0.208	0.178	0.152	0.130	0.112
11	0.543	0.446	0.368	0.305	0.253	0.211	0.176	0.148	0.125	0.105	0.089
12	0.512	0.413	0.334	0.272	0.222	0.182	0.450	0.123	0.102	0.085	0.071
13	0.483	0.382	0.304	0.243	0.195	0.157	o.127	0.103	0.084	0.068	0.056
14	0.456	0.354	0.276	0.217	0.171	0.135	0.107	0.086	0.069	0.055	0.044
15	0.430	0.328	0.251	0.194	0.150	0.117	0.091	0.071	0.056	0.044	0.035

In order to show how this analysis might be applied, the following hypothetical example will compare two compressors which are alternative pieces of equipment being considered for a new factory. The equipment costs came from actual quotations. Some of the other costs are actual and some assumed on rational bases, all for the year 1991. The two alternatives will be called compressor A and compressor B. The above analytical method will be applied to their comparison.

Electric power costs: 13.0¢ per kilowatt-hour
Cost of capital: 9%
Cost of oil and oil disposal, $6.50 per gallon
Cost of water and water disposal: $3.28 per 1,000 gals.
Compressor capacity (as per Section 4): 525 scfm
Shifts worked: 2
Hours of operation per year: 4450

Full load, 40%	2225 hours
Half load, 50%	1780 hours
One-tenth load, 10%:	445 hours

Inflation, over compressor life: 3.5% per year
Life of both compressors: 10 years
Environmental and operational factors: none
     favoring one compressor over the other

For depreciation, the U.S. Tax Codes as of 1991 set the following depreciation schedule for such equipment:

Ownership Year	Depreciation %
1	14.29
2	24.49
3	17.49
4	12.49
5	8.93
6	8.92
7	8.93
8	4.46

These schedules are changed from time to time by the U.S. Congress. The current or "soon to be expected" depreciation schedule, with appropriate investment tax credit, should be used. For this example assume a combined federal and state marginal tax rate of 38.6 percent and no investment tax credit, since there is none in the law as of 1991.

Additional data can be developed on each compressor as follows:

	Comp.A	Comp.B
Lubrication	Oil	Oil
Horsepower	125	100
Capacity, scfm	565	565
Horsepower required, full load	124	109
Horsepower required, half load	70	57
Horsepower, one-tenth load	33	18
Oil required, gals/mo	6	6
Water required (gals/minute full load)	9.0	8.5
Maintenance cost, per hp/mo	$2.50	$2.50
Initial costs (purchase + installation)	$34,000	$48,000
Salvage value	$4,760	$4,800
Electric motor efficiency, full load	0.92	0.92

It should be noted that partial load performance must be considered as it seriously effects power costs. On the above bases then, the operating costs can be computed for tabulation. The first year costs are shown on the next page.

These costs can be projected into future years by applying the rate of inflation, 3.5 percent, to each successive year. All of the information is now ready for tabulation as shown on Tables 5-2 and 5-3.

These tables are the same except that one describes total costs of compressor A and the other, the total costs of compressor B.

The first column, of course, is the year of life or use of the compressor.

Compressors

POWER COST:	A	B
Full load (hours × hp/motor eff. × 0.7455)	223,569	196,524
for example,(2,225 × 124/0.92 × 0.7455) for Comp.A		
Half load (hours × hp/motor eff. × 0.7455)	100,967	82,216
One-tenth load (hours × hp/motor eff. × 0.7455)	11,900	6,491
Total power, kWh	336,435	285,231
Total power cost (total power × cost per kWh)	$ 43,737	$37,080
OIL COST:		
Total gallons	6	6
Cost (total gallons × cost per gallon)	$39.00	$39.00
WATER COST		
Full load (minutes × % full power x water use per min.)	1,201,500	1,134,750
for example, (2,225 × 60 × 124/124 × 9) for Comp.A		
Half load (minutes × % full power × water use per min.)	542,613	474,721
One-tenth load (min × % full power × water per min.)	63,951	37,478
Total water use, gals	1,808,064	1,646,949
Total water cost	$5,930	$5,402
MAINTENANCE:		
Total cost (Cost/hp/mo × hp × 12)	$3,720	$3,270
For example, (2.50 × 124 × 12) for Comp. A		
TOTAL FIRST YEAR OPERATING COST:		
Total cost (power + oil + water + maintenance)	$53,426	$45,791

The second column includes the capital costs, initial and salvage.

The next column includes all operating costs, energy, oil, maintenance and cooling water, all as calculated above and adjusted for inflation.

The fourth column includes the rate of depreciation for standard eight-year depreciation (check with a tax accountant to be sure which depreciation schedule you should use in your industry).

The fifth column is the product of the depreciation rate and the initial compressor cost (col. 4 × col. 2, first year). This product is the amount which company management will report on tax returns.

The sixth column contains the tax savings due to depreciation. It is the product of the depreciation and the applicable tax rate, combined federal and local (0.386 × col. 5). If future tax laws provide an investment tax credit (itc), the savings should be added here for the year or years in which savings are allowed. Simply add the savings due to depreciation to the savings due to the tax credit and enter into this column.

The seventh column is the total of all cash outlays for the year (col. b + col. c - col. f).

The eighth column is the discount factor to present value, which is taken from Table 5-1.

The last column is the product of each year's cash outlay and the discount factor (col. g x col. h). A total of this column, as shown at the bottom of the table, is the total lifetime cost of the equipment and its operation, factored to present value.

Table 5-2 Present Value Lifetime Cost Projection
Compressor A

Year (a)	Initial & Salvage Costs $ (b)	Operating Costs $ (c)	Rate of Depreciation (d)	Depreciation for Year $ (e)	Tax Saving Depreciation (+ itc*) (f)	Cash outlay for Year $ (g)	Discount Factor to Present Value (h)	Present Value of Cost Per Year $ (j)
1	$34,000	$53,426	14.29%	$4,859	$1,875	$85,551	0.959	$82,043
2		$55,296	24.49%	$8,327	$3,214	$52,082	0.880	$45,832
3		$57,231	17.49%	$5,947	$2,295	$54,936	0.808	$44,388
4		$59,234	12.49%	$4,247	$1,639	$57,595	0.741	$42,678
5		$61,308	8.93%	$3,036	$1,172	$60,136	0.680	$40,892
6		$63,453	8.92%	$3,033	$1,171	$62,283	0.624	$38,864
7		$65,674	8.93%	$3,036	$1,172	$64,502	0.573	$36,960
8		$67,973	4.46%	$1,516	$585	$67,387	0.526	$35,446
9		$70,352				$70,352	0.483	$33,980
10	($4,760)	$72,814				$68,054	0.444	$30,216
Computation:		Inflation Added	Tax Law	dxb (1st yr.)	(tax rate)e+itc	b + c - f	Table 5-1	gxh

*Investment tax credit (itc) is available only in some years at the whim of Congress; refer to current income tax provisions.

Total Present Value, All Costs (j) $431,300

Table 5-3 Present Value Lifetime Cost Projection
Compressor B

Year (a)	Initial & Salvage Costs $ (b)	Operating Costs $ (c)	Rate of Depreciation (d)	Depreciation for Year $ (e)	Tax Saving, Depreciation (+ itc*) (f)	Cash outlay for Year $ (g)	Discount Factor to Present Value (h)	Present Value of Cost Per Year $(l)
1	$48,000	$45,791	14.29%	$ 6859	$2,648	$91,143	0.959	$87,406
2		$47,394	24.49%	$11,755	$4,538	$42,856	0.880	$37,713
3		$49,052	17.49%	$ 8,395	$3,241	$45,812	0.808	$37,016
4		$50,769	12.49%	$ 5,995	$2,314	$48,455	0.741	$35,905
5		$52,546	8.93%	$ 4,286	$1,655	$50,892	0.680	$34,606
6		$54,385	8.92%	$ 4,282	$1,653	$52,733	0.624	$32,905
7		$56,289	8.93%	$ 4,286	$1,655	$54,634	0.573	$31,305
8		$58,259	4.46%	$ 2,141	$ 826	$57,433	0.526	$30,210
9		$60,298				$60,298	0.483	$29,124
10	($4.800)	$62.408				$57,608	0.444	$25,578
Computation:		Inflation Added	Tax Law	dxb (1st yr)	(tax rate)e+itc	b+c-f	Table 5-1	gxh

Total Present Value, All Costs (j) $381,770

*Investment tax credit (itc) is available only in some years at the whim of Congress; refer to current income tax provisions.

It can be seen that in this particular case, and with the assumptions made, that compressor B is the preferred purchase, even though it costs much more, initially, than compressor A.

Any analysis, such as this one, is only as good as the information upon which it is based. Great care should be exercised in making sure that these inputs are correct, and some of them may not always be easy to get. Equipment catalogs do not always make available both full-load and partial load energy consumption information, but this information is crucial and must be determined. Also, maintenance costs should be determined carefully, preferably from actual cost records and experience of maintenance personnel. The cost of money and the inflation rate are closely related. When one is high, so is the other. These should be estimated for the period of the life of the equipment. The cost of money varies greatly from industry to industry. It will be lower, for example, for utility companies with their reasonably well assured returns on investment than for some other higher risk and higher return enterprises.

If all of these steps are followed carefully, then investment decisions made in this manner are quite likely to be optimum. In order to assist the reader with subsequent calculations of this type, Table 5-1, the present value factor table, and the life-time cost projection, Table 5-4, in blank form, are included. These may be copied for use by the reader.

This analytical approach is useful for comparing alternate courses of action in a wide variety of situations, including investment, operations, property purchase and others. For example, many of the electric motor manufacturers have analyses available of their new high efficiency, and more expensive, motors. These companies also use present value analysis and some can offer computer generated estimates of motor lifetime savings. In most cases, the high efficiency motors are easily cost effective for two or more shifts of compressor operations. However, each case should be independently analyzed.

Table 5-4 Present Value Lifetime Cost Projection

Year (a)	Initial & Salvage Costs $ (b)	Operating Costs $ (c)	Rate of Deprecia-tion (d)	Deprecia-tion for Year $ (e)	Tax Saving, Depreciation (+ tax Credit*, (f)	Cash out-lay for Year $ (g)	Discount Factor to Present Value (h)	Present Value of Cost Per Year $ (j)
Computation:		Inflation Added	Tax Law	dxb (1st yr.)	0.49e+0.1b	b+c-f	Table 5-1	gxh

Total Present Value, All Costs (j) _____ .

*Investment tax credit is available only in some years at the whim of Congress; refer to current income tax provisions.

The basic data collected in support of these analyses can be used in other ways, as described below.

5.2.2 Cost of Compressed Air

To determine the cost of compressed air, the same basic information is used, but it is summarized differently. For example, consider the cost of depreciating and operating compressor B in the example above. In order to calculate the cost per 1000 cubic feet of compressed air, first determine how much compressed air will be generated in the course of the year.

Begin with the total capacity and multiply the number of minutes per year by capacity and the average percentage output for each load condition. Sum the products, yielding the total air compressed during the year.

Referring to the original problem assumptions then in Section 5.2.1, the operating profile is:

Capacity: 525 scfm
2225 operating hours at 100% capacity
1780 operating hours at 50% capacity
445 hours at 10% capacity.

This converts and permits calculation of output:

2225 hours at 525 scfm
1780 hours at 262. 5 scfm
445 hours at 52. 5 gcfm

$$2225 \times 525 \times 60 \quad = \quad 70.1 \times 10^6 \text{ scf}$$
$$1780 \times 262. 5 \times 60 \quad = \quad 28. 0 \times 10^6 \text{ scf}$$
$$445 \times 52. 5 \times 60 \quad = \quad \underline{1. 4 \times 10^6 \text{ scf}}$$
$$99. 5 \times 10^6 \text{ scf}$$

Assume a superbly maintained system with only 10 percent losses to leakage, blowdowns, etc.

$$99. 5 \times 0. 9 = 89.55 \qquad \text{Use } 90 \times 10^6 \text{ scf used}$$

Next develop costs, by year, using forecast of operating costs and depreciation from Table 5-3. The total cost for each year should be divided by the useful output, 90×10^3, to get cost per 1000 scf, as follows:

Year	Oper. Cost	Depreciation	Total Cost	Cost/1000 scf
1	$45,791	$ 6,859	$52,650	$0.59
2	$47,623	$11,755	$59,378	$0.66
3	$49,528	$ 8,395	$57,923	$0.64
4	$51,509	$ 5,995	$57,504	$0.64
5	$53,569	$ 4,286	$57,855	$0.64
6	$55,712	$ 4,282	$59,993	$0.67
7	$57,940	$ 4,286	$62,227	$0.69
8	$60,258	$ 2,141	$62,399	$0.69
9	$62,668		$62,668	$0.70
10	$65,175		$65,175	$0.72

10 Year Average: 0.66

The total analysis then has provided not only a decision about what kind of equipment to purchase, but also what compressed air costs may be expected over the life of the equipment. This can be useful in planning manufacturing operations and for computing the costs of pressure drop and leakage.

5.2.3 Pressure Drop and Leakage Cost Reduction

For computing maintenance schedules, only two additional steps need to be performed. First, determine two cost inputs: the cost of taking a maintenance action, and the rate of increase in operating costs between these maintenance actions.

Then use the simple formula:

$$t = \sqrt{2\frac{m}{r}} \quad \text{where:}$$

t = optimum time between maintenance actions

m = the cost of the maintenance action, $

r = rate of increase in operating costs between maintenance actions, $ per [unit of time]^2

5.2.3.1 Sample Pressure Drop Calculation

For example, in the example system described above, assume that a paper filter element costs $20.50 and replacing it requires 30 minutes of labor costing $50 per hour, including overhead. Between filter changes, the pressure drop across the filter increases at a rate of 0.8 psi per month. This increases the power required (from the compressor power curves) by $0.55 \times 0.8 = 0.44\%$ per month.

Since the annual operating costs all are power related, and total $45,791 (using compressor B Table 5-3), the rate of cost increase is:

$$r = \frac{45,791}{12} \times 0.0044 = \$16.79 \text{ per month per month}$$

m = filter cost + labor

=$20.50 + $25.00

= $45.50

then the optimum time between filter changes,

$$t = \sqrt{2\frac{m}{r}}$$

$$= \sqrt{2\frac{45.50}{16.79}} = 2.32 \text{ months}$$

or about every 10 weeks.

If there is a pressure difference alarm on the filter, it should be set to alarm at a pressure drop equal to the pressure drop across a new element plus

0.8 (psig per mo.) × 2.32 (mos.)

If the pressure drop across a new element is 2 psi, then the alarm setting should be

2 + (0.8 × 2.32) = 3.86 psig; use 3.9 or 4.0 psig

5.2.3.2 Leakage Cost Calculations
The same approach can be taken to leakage. Assume, for example, that a system check for leaks, with repairs as usually found to be needed, requires 8 hours of maintenance labor (2 men 1/2 day on Saturday mornings) at 1-1/2 times the base hourly rate. The labor cost is 8 × 50 × 1-1/2 = $600.00. Materials usually are required in the amount of $70.00. The total, m, is $670 00.

Also, it has been found that, if no leakage maintenance is done, the leakage rate increases each month by 5% of total air used. The cost of 5% of the total air used is at least

0.05 × 45,791 = $2290 per year
    or $191.00/mo.

r = $191. /month/month

Using the formula

$$t = \sqrt{2\frac{m}{r}} = \sqrt{2\frac{670}{191}} = 2.65 \text{ months} = 11 \text{ weeks}$$

The system should be maintained for leak reduction every two and a half months for optimum cost savings.

Several things should be remembered. First, make sure that all of the factors in the equations are in the same units. If the rates of cost increase are in dollars per month per month, then the time between maintenance will be in months. All costs should be in dollars.

Finally, remember that the above examples may or may not be typical of other factories or locations. Each system requirement is different and must be evaluated independently. For example, if the above calculations had been applied to compressor A with its higher energy and operating costs, then the minimum system maintenance action timing (t) would be reduced, calling for higher maintenance costs. This is still another reason for purchasing more energy efficient equipment.

Also, this type of calculation should not be the only basis for maintenance action. If a leak or excessive pressure drop is detected, it should be fixed immediately, rather than on a schedule.

6.0 SUMMARY AND CONCLUSIONS

The important steps in saving energy in compressed air systems are:

(1) Understanding the requirements
(2) Proper system design
(3) Well engineered system modifications
(4) Economic criteria for equipment selection
(5) Clean, leak-free and obstruction-free installation
(6) Efficient operations
(7) Good Maintenance
(8) Dollar conscious management of all aspects of compressed air systems

6.1 Understanding the Requirements

One important factor here is a proper factory layout to make sure everything is in the right place. The compressor(s) should be where it (they) can get clean, dry air from as cool a spot as possible, preferably the northwest or shady side of the factory. The heavy air-using operations should be as near to the compressor as possible. If the compressor(s) can be located so that it is (they are) not too far from a central position, that will be better. Engineering judgement and a comparison of the total cost of alternatives will be important in balancing these considerations with other operative and production requirements.

Next, the air requirements must be specified carefully as to quantity, pressure and air quality. Where the pressure and quality are not standard throughout the plant, then economic study must be made of the advisability of providing several different types of air: both dried and undried, high pressure and low pressure, etc. Localized treatment or compression for special operations frequently is economic and conserving of energy. Provision should be made for potential growth, noting most especially near-term management plans.

When the layout is complete, the immediate and potential requirements for compressed air defined, and the supply air characteristics determined, then the distribution system can be defined and, finally, the compression and conditioning equipment can be specified.

6.2 Proper System Design

By the time all of the above has been determined, then enough information will exist to permit the design of the distribution system. In view of the potential for energy loss through pressure drop incident to such systems, the best of engineering design practice should be used in the design of the system. Oversizing is preferred to undersizing. Time and money might be saved by applying the newer computer technology to this necessarily detailed design task, but the calculations are not complex.

6.3 System Modifications

The same careful approach should be made to system modification. The requirement should be understood thoroughly and each modification designed in accordance with modern system design practice as per 6.2 above. In this way, the problem of overloaded branch lines and headers can be avoided, alleviating the incident excessive pressure drops and lost efficiency. Every time an addition or modification is made, its effect on the entire system should be considered. The operating cost of overloaded sub-headers or branch lines can be higher than realized because of the reduced energy content of the lower pressure air. This tends to increase production labor costs and induce greater air usage than necessary, with even further pressure drop and efficiency loss and possible catastrophic failure of a portion of the system.

6.4 Economic Criteria for Equipment Selection

The equipment with the lowest purchase price frequently is not the least expensive. The energy and other operating costs over equipment lifetime can easily, vastly exceed the purchase price. The best approach is a present value analysis of all costs over the lifetime of the equipment being considered. Care should be taken to match the equipment to the requirements. Too large a compressor operating at partial loads is less efficient; operating the system at too high a pressure is equally wasteful; overdrying the air also is costly. High efficiency motors usually are cost effective for two or more shifts of operation, sometimes for one to one and a half shifts. Always determine energy requirements of all equipment at full load, partial load and unloaded condition. This information always is worth acquiring, even though sometimes it is difficult to get. Equipment maintenance costs, also, are important, but normally they are small compared to power costs.

6.5 Clean Installation

Make sure that initial installation is inspected to be leak-free and free of obstructions. Hurried installations frequently leave behind leaky joints. Air lines do not immediately leak water onto the floor so the leaks can go undetected, sometimes for years. Only proper construction supervision can assure a good installation. Debris left behind in the distribution piping frequently will appear later as serious maintenance problems in traps, tools and other pneumatic devices. Accumulated dirt will build-up at fittings and obstruct lines.

6.6 Efficient Operations

Do not operate the system when manufacturing is not taking place. Explore shutting down even at lunchtime. Nighttime operations almost never can be justified unless there is a manufacturing nightshift. If it is necessary that instrument or other air availability be maintained at all times, then a separate, smaller, supply should be installed for these requirements. In multiple compressor operation, avoid operating more compressors than are appropriate to current manufacturing requirements. Consider alternate, more energy efficient, designs and tooling systems.

6.7 Good Maintenance

Maintenance is relatively important to compressed air systems as compared to other types of systems used in industry. Almost every part of the system is subject to a deterioration of performance through poor maintenance. Almost every deterioration of performance results in an increase in power consumption, either directly or indirectly, and a commensurate increase in operating cost. The worst problem of all is leakage. Leaks seem to spring forth of their own volition and subduing them is a constant battle. Seeking out leaks should be done on a regular basis.

Contamination, system modifications and improper sizing and application of fittings and hoses can result in excessive pressure drops. Pressure should be monitored at work stations, as should pressure drops in other key areas such as filters, the compressor inlet and the distribution system.

Compressor performance should be checked regularly. A well maintained compressor will use less power than will one that is not receiving proper care.

6.8 Dollar Conscious Management

Something as valuable and useful as compressed air should be managed with wisdom and as much economy as can be practiced. Therefore, a cost-consciousness in approaching all aspects of compressed air is indicated. The frequency of routine maintenance practices can be cost-optimized by modern cost analysis. The same can be said for equipment selection and distribution system layout. All factory managements should be aware of the lifetime costs of their air systems, of which energy cost is the greatest single element.

This utility deserves careful attention at the highest levels of factory management.

APPENDICES

Appendix A

BIBLIOGRAPHY

1. *Air Compressor Evaluation.* Worthington Compressors, 1976

2. *Air Compressor Installation and Operating Instructions.* Curtis-Toledo, January 1978.

3. *Air Compressor Selection and Application 1/4 HP Through 25 HP.* Compressed Air Gas Institute, 1980.

4. *Air Compressor Workshop.* Elliott Company, United Technologies.

5. Anderson, Michael E. "Tips on Gaining Maximum Service Life From a Compressor," *Chemical Equipment*, August 1982.

6. Argonne National Laboratory. *Classification and Evaluation of Electric Motors and Pumps.* NTIS, 1980.

7. Bein, Thomas W. *Foundations For Computer Simulation of a Low Pressure Oil Flooded Single Screw Air Compressor.* DTNSRDC, 1981.

8. *Better Air Power System.* A. Ingersoll-Rand, 1973.

9. "Blowoff Jet-Energy-Savings-Opportunity." *Compressed Air Magazine*, 1978.

10. *Compressed Air & Gas Drying*. Compressed Air and Gas Institute.

11. *Compressed Air System Survey*. Industrial Division, Schramm, Inc., Bulletin SSB-578

12. *Compressor Selector for Industry (CSI)*. WGI-127 Worthington Corporation, 1975.

13. *Condensed Air Power Data*. Ingersoll-Rand, 1978.

14. Cowern, E. H. "Application Guidelines for Energy Efficient Motors." *Hydraulics & Pneumatics*, March 1982, pp. 148-154.

15. *Comparative Study of the Energy Characteristics of Powered Hand Tools*. Prepared for the U.S. Department of Energy, NTIS SAN-1731-Tl and SAN1731-T2.

16. Ellis, William D. *Compressed Air and Gas: In Manufacturing*. Compressed Air and Gas Institute, 1976.

17. Ellis, William D. *Compressed Air and Gas in Industry*. Compressed Air and Gas Institute, 1980.

18. *Energy Efficiency and Electric Motors*. Prepared for U.S. Department of Energy, HCP/M50217-01, reprinted 1978.

19. Engineering Department of Crane. *Flow of Fluids Through Valves, Fittings, and Pipe*. Crane Co., 1969.

20. Foss, R. Scott. "Fundamentals of Compressed Air Systems." *Plant Engineering*, May, June, July 1981.

21. Gibbs, Charles W. *Compressed Air and Gas Data*. Ingersoll-Rand Company, 1981.

22. *Hankison Compressed Air Dryers*. Hankison Bulletin DB-100-1.

23. Henderson, C. "A Primer for Air Dryer Selection." *Hydraulics & Pneumatics*, November 1980.

24. Henderson, Charles. "The Economics of Air Drying." *Hydraulics & Pneumatics*, October 1981.

25. "How to Choose the Right Compressor." *Compressed Air*. 1981; reprinted from *Compressed Air*, 1978.

26. Ingersoll-Rand Company. *Energy System Facts & Data*.

27. Iwama. N., et al. *A Microprocessor-Based Compressed Air Supply System*. IECI. 1979 Proceedings.

28. Luther, Ward. "Determining Centrifugal Compressor Performance." *Power*, March 1979.

29. McCormick, J.A., ed. "Industrial Energy Conservation." *Compressed Air*, March 1980.

30. McCullough, John E. and Fritz Hirschfeld. "The Scroll Machine—." *Mechanical Engineering*, 1979.

31. Miller, William S. *Machine Design—1982 Electrical and Electronics Reference Issue*. U.S.A., May 1982.

32. Montgomery, David C. *How to Specify and Evaluate Energy-Efficient Motors*. General Electric, December 1981.

33. O'Keefe, William. "Air Compressors." *Power*, June 1977.

34. O'Keefe, William. "Compressed Air Auxiliary Equipment." *Power*, December 1978.

35. O'Neil, F.W. "Compressed Air Data." *Compressed Air Magazine*. Ingersoll Rand Company, 1955.

36. Oviatt, Mark D. and Richard K. Miller. *Industrial Pneumatic Systems, Noise Control and Energy Conservation*. Fairmont Press, Inc., 1981.

37. "RDL for Oil-Free Air." *Compressed Air Magazine*, November 1982.

38. Reason, John. "AC Motor Control." *Power*, February 1981.

39. Robson, N.H. "Drain The Traps and Save The Compressor," *Hydraulics & Pneumatics*, March 1982, pp 89-90.

40. Rollins, John P. *Compressed Air and Gas Handbook*. Compressed Air and Gas Institute, 1973.

41. *Stationary Compressor Market, The*. Frost & Sullivan, Inc. Report #720, June 1979.

42. Stull, N.R., et al. *Air Compressors Conditioning, Costs and the Crunch*. Penton/IPC, 1978.

43. *Westinghouse Training Course—Motor and Control*. Westinghouse Electric Corporation. September 1981.

Appendix B

FREQUENTLY USED ABBREVIATIONS

A	- Amperes
ac	- alternating current
ASME	- American Society of Mechanical Engineers
bhp	- brake horsepower
Btu	- British thermal units
°C	- degrees Celsius
cfm	- cubic feet per minute
cm	- centimeters
D	- diameter
dc	- direct current
Δp	- pressure differential
°F	- degrees Fahrenheit
FL	- full load
ft, '	- feet
ϕ, ϕ	- phase
g	- grams
gpm	- gallons per minute
hc	- hydrocarbons
hp, HP	- horsepower
H_2O	- water
HVAC	- Heating, Ventilating and Air Conditioning
Hz	- Hertz
I	- current
icfm	- cubic feet per minute at inlet conditions
in., "	- inches

kVA	- kilovolt-Amperes
kVAr	- kilovolt-Amperes, reactive
kW	- kilowatts
kWh	- kilowatt hours
L	- length
lb	- pounds
L/D	- equivalent length, in diameters
mo	- month
μ	- microns
NEMA	- National Electrical Manufacturer's Association
PD	- Positive displacement
pdp	- pressure dew point
pf	- power factor
ppm	- parts per million
psi	- pounds per square inch
psia	- pounds per square inch, absolute
psid	- pressure drop, pounds per square inch
psig	- pounds per square inch above atmospheric pressure
R	- resistance
rh	- relative humidity
rms	- root mean square
rpm	- revolutions per minute
scf	- standard cubic feet
scfm	- standard cubic feet per minute
scr	- silicon controlled rectifier
V	- Volts
VA	- Volt-Amperes
W	- Watts
w/	- with

APPENDIX C

EXTRA TABLES

The following pages may be removed for use by engineering personnel.

Air User Requirements — 6 copies

Present Value Tables (mid year) — 6 copies

Present Value Lifetime Cost Projections — 6 copies

AIR USER REQUIREMENTS

Process/ Branch #	Drop#/ Air User	Flow (icfm)	Min. Press. (psig)	Max. Temp. (°F)	Water Content	Oil Content

AIR USER REQUIREMENTS

Process/ Branch #	Drop#/ Air User	Flow (icfm)	Min. Press. (psig)	Max. Temp. (°F)	Water Content	Oil Content

AIR USER REQUIREMENTS

Process/ Branch #	Drop#/ Air User	Flow (icfm)	Min. Press. (psig)	Max. Temp. (°F)	Water Content	Oil Content

AIR USER REQUIREMENTS

Process / Branch #	Drop# / Air User	Flow (icfm)	Min. Press. (psig)	Max. Temp. (°F)	Water Content	Oil Content

AIR USER REQUIREMENTS

Process/ Branch #	Drop#/ Air User	Flow (icfm)	Min. Press. (psig)	Max. Temp. (°F)	Water Content	Oil Content

AIR USER REQUIREMENTS

Process / Branch #	Drop# / Air User	Flow (icfm)	Min. Press. (psig)	Max. Temp. (°F)	Water Content	Oil Content

PRESENT VALUE FACTOR TABLE
For Costs Incurred Throughout Each Year
(Mid-Year Present Values)

Discount Rate or Cost of Capital

Year	6%	8%	10%	12%	14%	16%	18%	20%	22%	24%	26%
1	0.972	0.963	0.955	0.946	0.939	0.931	0.924	0.912	0.910	0.903	0.897
2	0.917	0.892	0.868	0.845	0.823	0.803	0.783	0.764	0.745	0.728	0.712
3	0.864	0.826	0.789	0.754	0.722	0.629	0.663	0.637	0.611	0.587	0.565
4	0.816	0.764	0.717	0.674	0.634	0.596	0.562	0.530	0.501	0.474	0.448
5	0.770	0.708	0.652	0.601	0.556	0.514	0.476	0.442	0.411	0.382	0.356
6	0.726	0.655	0.592	0.537	0.487	0.443	0.404	0.368	0.337	0.308	0.282
7	0.685	0.607	0.539	0.479	0.428	0.382	0.342	0.307	0.276	0.248	0.224
8	0.646	0.562	0.490	0.428	0.375	0.329	0.290	0.256	0.226	0.200	0.178
9	0.610	0.520	0.445	0.382	0.329	0.284	0.246	0.213	0.185	0.162	0.141
10	0.575	0.482	0.405	0.341	0.289	0.245	0.208	0.178	0.152	0.130	0.112
11	0.543	0.446	0.368	0.305	0.253	0.211	0.176	0.148	0.125	0.105	0.089
12	0.512	0.413	0.334	0.272	0.222	0.182	0.450	0.123	0.102	0.085	0.071
13	0.483	0.382	0.304	0.243	0.195	0.157	o.127	0.103	0.084	0.068	0.056
14	0.456	0.354	0.276	0.217	0.171	0.135	0.107	0.086	0.069	0.055	0.044
15	0.430	0.328	0.251	0.194	0.150	0.117	0.091	0.071	0.056	0.044	0.035

PRESENT VALUE FACTOR TABLE
For Costs Incurred Throughout Each Year
(Mid-Year Present Values)

Discount Rate or Cost of Capital

Year	6%	8%	10%	12%	14%	16%	18%	20%	22%	24%	26%
1	0.972	0.963	0.955	0.946	0.939	0.931	0.924	0.912	0.910	0.903	0.897
2	0.917	0.892	0.868	0.845	0.823	0.803	0.783	0.764	0.745	0.728	0.712
3	0.864	0.826	0.789	0.754	0.722	0.629	0.663	0.637	0.611	0.587	0.565
4	0.816	0.764	0.717	0.674	0.634	0.596	0.562	0.530	0.501	0.474	0.448
5	0.770	0.708	0.652	0.601	0.556	0.514	0.476	0.442	0.411	0.382	0.356
6	0.726	0.655	0.592	0.537	0.487	0.443	0.404	0.368	0.337	0.308	0.282
7	0.685	0.607	0.539	0.479	0.428	0.382	0.342	0.307	0.276	0.248	0.224
8	0.646	0.562	0.490	0.428	0.375	0.329	0.290	0.256	0.226	0.200	0.178
9	0.610	0.520	0.445	0.382	0.329	0.284	0.246	0.213	0.185	0.162	0.141
10	0.575	0.482	0.405	0.341	0.289	0.245	0.208	0.178	0.152	0.130	0.112
11	0.543	0.446	0.368	0.305	0.253	0.211	0.176	0.148	0.125	0.105	0.089
12	0.512	0.413	0.334	0.272	0.222	0.182	0.450	0.123	0.102	0.085	0.071
13	0.483	0.382	0.304	0.243	0.195	0.157	o.127	0.103	0.084	0.068	0.056
14	0.456	0.354	0.276	0.217	0.171	0.135	0.107	0.086	0.069	0.055	0.044
15	0.430	0.328	0.251	0.194	0.150	0.117	0.091	0.071	0.056	0.044	0.035

PRESENT VALUE FACTOR TABLE
For Costs Incurred Throughout Each Year
(Mid-Year Present Values)

Discount Rate or Cost of Capital

Year	6%	8%	10%	12%	14%	16%	18%	20%	22%	24%	26%
1	0.972	0.963	0.955	0.946	0.939	0.931	0.924	0.912	0.910	0.903	0.897
2	0.917	0.892	0.868	0.845	0.823	0.803	0.783	0.764	0.745	0.728	0.712
3	0.864	0.826	0.789	0.754	0.722	0.629	0.663	0.637	0.611	0.587	0.565
4	0.816	0.764	0.717	0.674	0.634	0.596	0.562	0.530	0.501	0.474	0.448
5	0.770	0.708	0.652	0.601	0.556	0.514	0.476	0.442	0.411	0.382	0.356
6	0.726	0.655	0.592	0.537	0.487	0.443	0.404	0.368	0.337	0.308	0.282
7	0.685	0.607	0.539	0.479	0.428	0.382	0.342	0.307	0.276	0.248	0.224
8	0.646	0.562	0.490	0.428	0.375	0.329	0.290	0.256	0.226	0.200	0.178
9	0.610	0.520	0.445	0.382	0.329	0.284	0.246	0.213	0.185	0.162	0.141
10	0.575	0.482	0.405	0.341	0.289	0.245	0.208	0.178	0.152	0.130	0.112
11	0.543	0.446	0.368	0.305	0.253	0.211	0.176	0.148	0.125	0.105	0.089
12	0.512	0.413	0.334	0.272	0.222	0.182	0.450	0.123	0.102	0.085	0.071
13	0.483	0.382	0.304	0.243	0.195	0.157	0.127	0.103	0.084	0.068	0.056
14	0.456	0.354	0.276	0.217	0.171	0.135	0.107	0.086	0.069	0.055	0.044
15	0.430	0.328	0.251	0.194	0.150	0.117	0.091	0.071	0.056	0.044	0.035

PRESENT VALUE FACTOR TABLE
For Costs Incurred Throughout Each Year
(Mid-Year Present Values)

Discount Rate or Cost of Capital

Year	6%	8%	10%	12%	14%	16%	18%	20%	22%	24%	26%
1	0.972	0.963	0.955	0.946	0.939	0.931	0.924	0.912	0.910	0.903	0.897
2	0.917	0.892	0.868	0.845	0.823	0.803	0.783	0.764	0.745	0.728	0.712
3	0.864	0.826	0.789	0.754	0.722	0.629	0.663	0.637	0.611	0.587	0.565
4	0.816	0.764	0.717	0.674	0.634	0.596	0.562	0.530	0.501	0.474	0.448
5	0.770	0.708	0.652	0.601	0.556	0.514	0.476	0.442	0.411	0.382	0.356
6	0.726	0.655	0.592	0.537	0.487	0.443	0.404	0.368	0.337	0.308	0.282
7	0.685	0.607	0.539	0.479	0.428	0.382	0.342	0.307	0.276	0.248	0.224
8	0.646	0.562	0.490	0.428	0.375	0.329	0.290	0.256	0.226	0.200	0.178
9	0.610	0.520	0.445	0.382	0.329	0.284	0.246	0.213	0.185	0.162	0.141
10	0.575	0.482	0.405	0.341	0.289	0.245	0.208	0.178	0.152	0.130	0.112
11	0.543	0.446	0.368	0.305	0.253	0.211	0.176	0.148	0.125	0.105	0.089
12	0.512	0.413	0.334	0.272	0.222	0.182	0.450	0.123	0.102	0.085	0.071
13	0.483	0.382	0.304	0.243	0.195	0.157	o.127	0.103	0.084	0.068	0.056
14	0.456	0.354	0.276	0.217	0.171	0.135	0.107	0.086	0.069	0.055	0.044
15	0.430	0.328	0.251	0.194	0.150	0.117	0.091	0.071	0.056	0.044	0.035

PRESENT VALUE FACTOR TABLE
For Costs Incurred Throughout Each Year
(Mid-Year Present Values)

Discount Rate or Cost of Capital

Year	6%	8%	10%	12%	14%	16%	18%	20%	22%	24%	26%
1	0.972	0.963	0.955	0.946	0.939	0.931	0.924	0.912	0.910	0.903	0.897
2	0.917	0.892	0.868	0.845	0.823	0.803	0.783	0.764	0.745	0.728	0.712
3	0.864	0.826	0.789	0.754	0.722	0.629	0.663	0.637	0.611	0.587	0.565
4	0.816	0.764	0.717	0.674	0.634	0.596	0.562	0.530	0.501	0.474	0.448
5	0.770	0.708	0.652	0.601	0.556	0.514	0.476	0.442	0.411	0.382	0.356
6	0.726	0.655	0.592	0.537	0.487	0.443	0.404	0.368	0.337	0.308	0.282
7	0.685	0.607	0.539	0.479	0.428	0.382	0.342	0.307	0.276	0.248	0.224
8	0.646	0.562	0.490	0.428	0.375	0.329	0.290	0.256	0.226	0.200	0.178
9	0.610	0.520	0.445	0.382	0.329	0.284	0.246	0.213	0.185	0.162	0.141
10	0.575	0.482	0.405	0.341	0.289	0.245	0.208	0.178	0.152	0.130	0.112
11	0.543	0.446	0.368	0.305	0.253	0.211	0.176	0.148	0.125	0.105	0.089
12	0.512	0.413	0.334	0.272	0.222	0.182	0.450	0.123	0.102	0.085	0.071
13	0.483	0.382	0.304	0.243	0.195	0.157	o.127	0.103	0.084	0.068	0.056
14	0.456	0.354	0.276	0.217	0.171	0.135	0.107	0.086	0.069	0.055	0.044
15	0.430	0.328	0.251	0.194	0.150	0.117	0.091	0.071	0.056	0.044	0.035

PRESENT VALUE FACTOR TABLE
For Costs Incurred Throughout Each Year
(Mid-Year Present Values)

Year	Discount Rate or Cost of Capital										
	6%	8%	10%	12%	14%	16%	18%	20%	22%	24%	26%
1	0.972	0.963	0.955	0.946	0.939	0.931	0.924	0.912	0.910	0.903	0.897
2	0.917	0.892	0.868	0.845	0.823	0.803	0.783	0.764	0.745	0.728	0.712
3	0.864	0.826	0.789	0.754	0.722	0.629	0.663	0.637	0.611	0.587	0.565
4	0.816	0.764	0.717	0.674	0.634	0.596	0.562	0.530	0.501	0.474	0.448
5	0.770	0.708	0.652	0.601	0.556	0.514	0.476	0.442	0.411	0.382	0.356
6	0.726	0.655	0.592	0.537	0.487	0.443	0.404	0.368	0.337	0.308	0.282
7	0.685	0.607	0.539	0.479	0.428	0.382	0.342	0.307	0.276	0.248	0.224
8	0.646	0.562	0.490	0.428	0.375	0.329	0.290	0.256	0.226	0.200	0.178
9	0.610	0.520	0.445	0.382	0.329	0.284	0.246	0.213	0.185	0.162	0.141
10	0.575	0.482	0.405	0.341	0.289	0.245	0.208	0.178	0.152	0.130	0.112
11	0.543	0.446	0.368	0.305	0.253	0.211	0.176	0.148	0.125	0.105	0.089
12	0.512	0.413	0.334	0.272	0.222	0.182	0.450	0.123	0.102	0.085	0.071
13	0.483	0.382	0.304	0.243	0.195	0.157	0.127	0.103	0.084	0.068	0.056
14	0.456	0.354	0.276	0.217	0.171	0.135	0.107	0.086	0.069	0.055	0.044
15	0.430	0.328	0.251	0.194	0.150	0.117	0.091	0.071	0.056	0.044	0.035

Present Value Lifetime Cost Projection

Year (a)	Initial & Salvage Costs $ (b)	Operating Costs $ (c)	Rate of Depreciation (d)	Depreciation for Year $ (e)	Tax Saving, Depreciation (+ itc*) (f)	Cash outlay for Year $ (g)	Discount Factor to Present Value (h)	Present Value of Cost Per Year $ (j)
Computation:	Inflation Added		Tax Law	dxb (1st yr.)	(tax rate)e+itc	b + c - f	Table 5-1	gxh

Total Present Value, All Costs (j)

*Investment tax credit (itc) is available only in some years at the whim of Congress; refer to current income tax provisions.

Present Value Lifetime Cost Projection

Year (a)	Initial & Salvage Costs $ (b)	Operating Costs $ (c)	Rate of Depreciation (d)	Depreciation for Year $ (e)	Tax Saving, Depreciation (+ itc*) (f)	Cash outlay for Year $ (g)	Discount Factor to Present Value (h)	Present Value of Cost Per Year $ (j)
Computation:		Inflation Added	Tax Law	dxb (1st yr.)	(tax rate)e+itc	b + c - f	Table 5-1	gxh

Total Present Value, All Costs (j)

*Investment tax credit (itc) is available only in some years at the whim of Congress; refer to current income tax provisions.

Present Value Lifetime Cost Projection

Year (a)	Initial & Salvage Costs $ (b)	Operating Costs $ (c)	Rate of Depreciation (d)	Depreciation for Year $ (e)	Tax Saving, Depreciation (+ itc*) (f)	Cash outlay for Year $ (g)	Discount Factor to Present Value (h)	Present Value of Cost Per Year $ (j)
Computation:	Inflation Added		Tax Law	dxb (1st yr.)	(tax rate)e+itc	b + c - f	Table 5-1	gxh

Total Present Value, All Costs (j)

*Investment tax credit (itc) is available only in some years at the whim of Congress; refer to current income tax provisions.

Present Value Lifetime Cost Projection

Year (a)	Initial & Salvage Costs $ (b)	Operating Costs $ (c)	Rate of Depreciation (d)	Depreciation for Year $ (e)	Tax Saving, Depreciation (+ itc*) (f)	Cash outlay for Year $ (g)	Discount Factor to Present Value (h)	Present Value of Cost Per Year $ (j)
Computation:	Inflation Added		Tax Law	dxb (1st yr.)	(tax rate)e+itc	b + c - f	Table 5-1	gxh

Total Present Value, All Costs (j) _____

*Investment tax credit (itc) is available only in some years at the whim of Congress; refer to current income tax provisions.

Present Value Lifetime Cost Projection

Year (a)	Initial & Salvage Costs $ (b)	Operating Costs $ (c)	Rate of Depreciation (d)	Deprecia- tion for Year $ (e)	Tax Saving, Depreciation (+ itc*) (f)	Cash out- lay for Year $ (g)	Discount Factor to Present Value (h)	Present Value of Cost Per Year $ (j)
Computation:	Inflation Added		Tax Law	dxb (1st yr.)	(tax rate)e+itc	b + c - f	Table 5-1	gxh

Total Present Value, All Costs (j)

*Investment tax credit (itc) is available only in some years at the whim of Congress; refer to current income tax provisions.

Present Value Lifetime Cost Projection

Year (a)	Initial & Salvage Costs $ (b)	Operating Costs $ (c)	Rate of Depreciation (d)	Depreciation for Year $ (e)	Tax Saving, Depreciation (+ itc*) (f)	Cash outlay for Year $ (g)	Discount Factor to Present Value (h)	Present Value of Cost Per Year $ (j)
Computation:	Inflation Added	Tax Law		dxb (1st yr.)	(tax rate)e+itc	b + c - f	Table 5-1	gxh

Total Present Value, All Costs (j)

*Investment tax credit (itc) is available only in some years at the whim of Congress; refer to current income tax provisions.

INDEX